Fire Safety Engineering - Measures, Policies, and Applications

Edited by Chiara Bedon, Andrea Lucherini and Flavio Stochino

Published in London, United Kingdom

Fire Safety Engineering - Measures, Policies, and Applications
http://dx.doi.org/10.5772/intechopen.1004964
Edited by Chiara Bedon, Andrea Lucherini and Flavio Stochino

Contributors
Carlos Salema, Chiara Bedon, Demetris Nicolaides, Ioanna Giannopoulou, Ira Helsloot, Kate T. Q. Nguyen, Lorenzo Veronese, Nicola Cella, Rafael Caldeirinha, Rajeendra Godakandage, Riccardo Del Bello, Ron A. C. de Wit, Thomas W. Loh, Yu Wang

First published in London, United Kingdom, 2025 by IntechOpen
IntechOpen is the global imprint of INTECHOPEN LIMITED, registered in England and Wales, registration number: 11086078, 167-169 Great Portland Street, London, W1W 5PF, United Kingdom

For EU product safety concerns: IN TECH d.o.o., Prolaz Marije Krucifikse Kozulić 3, 51000 Rijeka, Croatia, info@intechopen.com or visit our website at intechopen.com.

British Library Cataloguing-in-Publication Data
A catalogue record for this book is available from the British Library

Fire Safety Engineering - Measures, Policies, and Applications
Edited by Chiara Bedon, Andrea Lucherini and Flavio Stochino
p. cm.
Print ISBN 978-0-85014-829-9
Online ISBN 978-0-85014-828-2
eBook (PDF) ISBN 978-0-85014-830-5

If disposing of this product, please recycle the paper responsibly.

Meet the editor

Chiara Bedon is an Associate Professor at the University of Trieste, Department of Engineering and Architecture, in Italy, where she's the principal instructor for 'Structural Analysis', 'Steel Structures' and 'Seismic Resistant Structures'. She is actively involved in international research projects and networks (JRC-ERNCIP, NATO-SPS, COST, etc.) with a primary focus on structural systems, extreme design actions, steel and glass, and is a member of national and international boards in charge of the elaboration of design standards. Listed as a "Top 2% Scientist" by Stanford and a "Top Italian Woman Scientist" in engineering, she has received several research grants for young scientists. She is the principal investigator for several projects, which have received funding of up to 1. 6 million euros in recent years.

Flavio Stochino is an Associate Professor of Theory and Design of Structures at the University of Cagliari, Italy, recognized among Stanford's top 2% of researchers worldwide. His work focuses on concrete technology, structural analysis, and the performance of materials under extreme conditions. He has co-authored numerous international publications, including research on sustainable construction and the rehabilitation of concrete structures. Beyond academia, he is a sports coach and project leader for the Unica Sailing Team, supporting young athletes in their academic and athletic development. Flavio is also actively engaged in promoting innovation through industry-academic collaborations and sustainability-focused initiatives.

Andrea Lucherini holds the position of Senior Researcher at the Slovenian National Building and Civil Engineering Institute (ZAG) and Assistant Professor at the University of Primorska (Slovenia). His field of expertise is construction materials engineering, with a specialization in fire safety science and engineering, and he has co-authored more than 50 peer-reviewed papers, book chapters, scientific conference contributions, and technical reports. His research interests primarily focus on the performance of advanced construction materials and systems in fire, material behaviour at elevated temperatures, advanced experimental fire methodologies, and fire dynamics for performance-based fire safety design. Andrea has also received multiple scientific awards and distinctions, including the 2024 SFPE 5 Under 35 Award, the 2024 ZAG Turnšek Junior Award, the MSCA Postdoctoral Fellowship, and the IAFSS Sheldon Tieszen Student Award.

Contents

Preface

Fire safety engineering denotes, in general terms, the well-known practice of designing and installing systems, building components and/or structures that are able to prevent and reduce the risk of fire, minimizing the hazard for people and property. As such, it involves applying scientific and engineering principles in different fields in the form of safety policy measures, technologies, and innovative strategies.

From a practical point of view, however, fire safety engineering tasks are much more articulated and differentiated. It is, in fact, a multidisciplinary field that applies – case by case – scientific and engineering principles. In the same way as specific measures and applications, there are thus many policies and bureaucratic aspects to account for, such as experimental protocols for fire performance assessment, standards, etc.

In such a wide discussion, among others, it can be recognized that all these technical aspects and phenomena mostly depend on two key features, namely the propagation of fire (and all the multiple related tasks) and the fire response of building and construction materials, which are responsible for many different resisting and damage mechanisms to address.

While there is a deep international technical knowledge on many of the constituent issues of fire safety engineering, there are still open questions to explore and assess, and the scientific research provides significant advancements in different ways. In this sense, the choice of a given construction material is not a simple mechanical assumption for satisfying appropriate structural safety levels in cold conditions but is necessarily correlated to the possible effects of fire and their reaction in the presence of such an extreme design action. Mitigating fire-related risk and fire-induced effects can take advantage of retardant building materials and components. Several of them are already in use, while various research studies focus on elaborating and developing new mitigation solutions. In parallel, technological innovation is of utmost importance in a list of fundamental activities for emergency management. At the same time, however, technological components in buildings may also represent a possible source of additional risk against fire and thus should be appropriately addressed and verified, both with experimental and virtual methods.

This book project collects five different research contributions that clearly explain the relevance and variety of fire safety engineering issues.

Safety measures are mandatory for fire protection, but they all have a cost. In the same way as other fields, specific considerations can thus help safety policymakers find optimal solutions. The study reported in Chapter 1 presents a (social) cost-benefit analysis (SCBA) in the fire safety domain. More in detail, the cost-benefit analysis is developed for a case-study configuration, representing an automatic alarm transmission for a public alarm system (OMS). The chapter concludes with a consideration of some specific themes in relation to SCBA, such as the so-called value of a statistical life (VOSL) regarding fire, presenting some relevant outcomes for future elaborations.

Chapter 2, as a perspective chapter, opens the discussion to the use of mobile radio emergency communications for large-scale wildfire fighting. A case-study example for Portugal is presented. Effective communication is, in fact, notoriously crucial for emergency response operations. New technologies can support and optimize the activities of fire safety policymakers and especially emergency units, such as firefighters. The discussed case-study example of the Portugal experience gives evidence of the key role of push-to-talk terrestrial systems in ensuring the safety and effectiveness of firefighting teams operating in remote and challenging terrains. The analysis demonstrates the limitations of current public and private communication networks, particularly in low-density population areas where they are most needed during emergencies. As such, the need for specialized land-mobile radio systems that can operate in low-frequency bands (e.g., 400 MHz) becomes paramount.

Following the first two chapters, the last contributions of this book are more oriented and focused on building performance in fire in relation to components and materials.

Chapter 3 is dedicated to an in-depth analysis of ventilated facades and their role for fire safety purposes in buildings. In particular, attention is given to open-state cavity barriers, which typically have a key role in fire conditions. They allow airflow through under normal circumstances and react to compartmentalize the cavity in order to retard the spread of fire in case of need. Besides, there are several open questions to address, which are pointed out in Chapter 3, especially regarding the suitability of existing test methods to experimentally assess the fire performance of these open-state cavity barriers for ventilated facades. The authors present a sound review of existing methods and protocols and open the discussion to elaborating and proposing new testing approaches and methodologies.

In Chapter 4, the authors present an experimental study that is carried out to assess the potential of geopolymer materials for buildings. The proposed investigation has a double goal: (i) the use of inorganic polymers from construction and demolition waste is verified in terms of thermal and mechanical properties, and at the same time, (ii) their potential for fire safety engineering purposes is experimentally addressed. Test results in fire conditions show that the newly developed material represents not only a sustainable solution for buildings but can also preserve its integrity and original mechanical properties for a rational time interval during fire exposure, making it quite an interesting solution to explore further in the construction sector.

Chapter 5 presents a numerical investigation of the fire performance assessment of building-integrated photovoltaic (BIPV) components for buildings. As shown, modern technologies are increasingly used in construction for many positive aspects, including energy and sustainability concerns. On the other side, BIPVs, in their literal form, represent a multidisciplinary component in building design. They are expected to provide minimum structural safety levels like other primary or secondary building components, and careful attention is required, especially under extreme design actions, such as fire. The numerical study reported in Chapter 5, in particular, focuses on the possibility and potential of numerical methods to capture the thermal shock occurrence of BIPV systems under fire and takes advantage of a literature full-scale experiment to validate modelling assumptions. Future studies and investigations will certainly benefit from these preliminary numerical insights.

All readers are invited to benefit from this international experience and discussion on fire safety engineering. In doing so, a special thanks goes to the contributing authors and the reviewers.

Chiara Bedon
Department of Engineering and Architecture,
University of Trieste,
Trieste, Italy

Flavio Stochino
Department of Civil, Environmental Engineering and Architecture,
University of Cagliari,
Cagliari, Italy

Andrea Lucherini
ZAG Fire Laboratory,
Logatec, Slovenia

Chapter 1

Policy Decisions in Fire Safety: Application of Cost-benefit Analysis

Ron A.C. de Wit and Ira Helsloot

Abstract

All safety measures have costs and benefits associated to them. A social cost-benefit analysis (SCBA) therefore ideally is a part of fire safety policy decision-making. This chapter introduces the concept of SCBA and its constituting elements. A CBA case study on automatic fire alarm systems is presented to show how an SCBA can be computed. In this case study, the safety benefits of these systems do not outweigh the costs. The SCBA was basis for a policy decision to limit the number of objects for which an automatic fire alarm system is obligatory. Part of the input for any SCBA is insight into the value that the public attributes to aspects such as the presence of the fire service or the value of lives saved by fire safety measures. For the aspect of 'lives saved' in the context of fire safety, an estimate of the value of a statistical life (VOSL) is relevant. Not much research has been done to get insight into these values specifically for fire safety. We report on the outcome of an experiment to obtain these values in the Dutch context. Interesting enough, the fire VOSL was found to be a magnitude higher than the road safety VOSL.

Keywords: cost-benefit analysis, value of statistical life, fire safety, automatic fire alarm, safety benefits

1. Introduction

(Social) cost-benefit analyses (SCBAs) are worldwide used when making trade-offs and decisions about investments, projects or policy measures. This applies to the safety domain as well: policy makers here also have to decide upon investments in policies on all possible kind of risks. SCBAs are therefore, for example, regularly used in road safety policies [1], crime prevention policies [2] and health care [3]. In the fire safety domain, the application of SCBA is still relatively limited. Nevertheless, a series of reports and studies have also been made in this domain [4–8].

Our starting point is that policy decisions on fire safety at least require well-founded considerations on the deployment of public resources. Insight into social return means, among other things, insight into costs and revenues (benefits) of the various activities and measures regarding fire safety. By the way: 'more social return' might suggest that the current return is known. This is precisely one of the tricky points when applying SCBAs for the fire safety: the effect and exact yield (and value) of the activities in the current situation (the so-called null alternative) are more often not known than well known.

The purpose of this chapter is to provide an introduction to the use of the social cost-benefit analysis tool in the fire safety domain. As we will show, this is a tool that is becoming increasingly important also in this field of work. The chapter will provide the interested reader with a number of entry points for further in-depth study of the subject. This means that the SCBA instrument itself is briefly discussed, and we will focus on the (possibilities for) application within the fire safety domain. We present an elaborate case study so the reader may get an understanding of how to perform an SCBA him- or herself. The case study is a cost-benefit analysis of automatic alarm transmission in the public alarm system (OMS). The chapter concludes with a consideration of some specific themes in relation to SCBA such as the so-called value of a statistical life (VOSL) in relation to fire.

2. What is a cost-benefit analysis?

A cost-benefit analysis is a method that compares the expected costs of a particular issue or problem with the expected benefits. The aim is to choose the most advantageous solution in order to justify policy choices. Cost-benefit analysis is thus part of a broader set of economic evaluation methods aimed at policy cost-effectiveness analysis. In a 'good' CBA, not only the direct financial costs and benefits are included but also the indirect economic costs and benefits.

We speak about social cost-benefit analysis when we weigh up all current and future social benefits and drawbacks of a project by expressing them (as much as possible) in monetary terms. When the sum of all advantages (benefits) outweighs the sum of all disadvantages (costs), a project is socially justifiable.

An SCBA may reflect costs and benefits that cannot be directly expressed in monetary terms (e.g. a risk reduction). In that case, a financial translation should take place. In other words, the effects of the project or intervention should be monetised. Section 5.4 discusses this further.

A (social) cost-benefit analysis can be carried out in various ways. In general, the following steps can be distinguished:

- Problem analysis and demarcation: what is the problem and which items are relevant to include in the remainder of the analysis?

- Describe null alternative: what will happen if the existing situation is continued unchanged? This helps clarify what effect is attributable to any new measure.

- Identifying and quantifying impacts: this is generally the most difficult step in an (S)CBA.

- Visualising costs and monetizing effects: making an overview of balance costs and benefits.

- Conclusions and advice including variants and sensitivity analysis.

For a detailed explanation of the above steps and the SCBA methodology in general, see for example [4, 9, 10]. See Ref. [11] for specific SCBA application in the fire safety domain.

Conducting an SCBA is not an end in itself. An SCBA is a tool that can indicate whether or not an investment in a particular project is socially the right choice. It can also substantiate which project alternative is best or prioritisation can take place. By providing this information, the SCBA supports political-administrative decision-making. An SCBA helps prevent economically inefficient choices being made or taxpayers' money not being spent optimally. Helsloot [12] notes that 'in fact, little is still known about the essential cost-benefit questions that need to be answered in the various phases of making safety policies'. However, given the ambition to strive for 'more social return', this is an important prerequisite for underpinning decision-making within the physical safety domain.

The American public administration expert Sunstein reached similar conclusions in his research on risk policy: risk appreciation and risk assessment by society is influenced by a number of cognitive mechanisms that lead to 'biases' [13]. Examples include 'probability neglect', whereby judgements are made mainly on the effect and not on the (often small) probability, and the 'availability bias', whereby having or not having recent experience on hand leads to a higher or lower risk assessment.

This can lead to incorrect use of public resources in regulating risks. A (social) cost-benefit analysis helps put risks in the right perspective. Incidentally, this does not mean that the outcome of the cost-benefit analysis determines where to regulate and where not to regulate. That is and remains a political-administrative consideration but after the SCBA a well-informed one. Without a clear understanding of costs and benefits, 'risk regulation' too often becomes a 'stab in the dark', according to Sunstein [13].

3. Possible applications within the fire safety domain

Decision-making on investments in policies or measures for fire safety and firefighting takes place in different environments. In the context of public administration, political-administrative considerations take place, and a social cost-benefit analysis can provide rational arguments to support or weigh up a decision. This may involve considerations within the firefighting domain (e.g. investing in more fire prevention education versus improving turnout times with an extra fire station) or between domains (e.g. traffic safety versus fire safety).

For an entrepreneur, a decision on an investment in a fire safety measure may also involve a cost-benefit analysis (e.g. whether or not to instal a sprinkler in relation to the [additional] cost of the insurance premium) or business risk [4]. In principle, the SCBA tool can be applied to any of the tasks of the fire service. The **Table 1** gives a qualitative overview of the (im)possibilities for applying CBA for the various task areas.

In practice, the use of SCBA in the fire safety domain is rather limited. In the Dutch context, the following examples can be given. In the Dutch Safety Regions Decree, application of SCBA is mentioned for (only) one task specifically; that is, in the explanatory memorandum, the following is stated under the heading 'Response times for basic firefighting services': 'The organisation of firefighting services is the result of an administrative cost-benefit assessment based on a fire risk assessment as part of the risk profile that the board of the safety region establishes pursuant to Section 15 of the Act'. In practice, this SCBA is never made explicit.

Theme		Possible purpose (S) CBA	Costs	Impact/benefits	Remark
Fire safety	Policies aimed at increasing safety awareness (such as information & education)	Visualising returns on existing or new policies	• Staff (s.a. pay, training) • Material	Behavioural influence should lead to lower risk of fire. As a result, fewer casualties and/or damage.	Limited view of baseline alternative. No data on relationship effort and effect. At most theoretical estimate.
	Advise to public authority (building codes, spatial planning, environment) and supervision	Visualising returns on existing policies or intensification of policies	Staff (s.a. pay, training)	Permit requirements and controls with penalties leading to measures and provisions. As a result, less risk of fire and possibly more limited consequences. As a result, fewer casualties and/or damage.	Limited view on baseline alternative. Fire damage is known. Not known is how much damage has been prevented. Relationship between more rules/checks and fewer casualties or injuries is unknown.
	Safety measures such as residential sprinklers	Decision on a policy on regulation of residential sprinklers	Investment in installation	More rapid control of fires. Fewer casualties and/or damage.	Data needed to quantify relation between measure and effect.
Incident response	Basic fire service	Coverage plan, (re)orientation on stations (keep, build new or close).	• Staff (including pay, training) • Equipment, material • Buildings • Traffic unsafety high priority turnout	Preparedness, 'being there' and faster turnout times. As a result, fewer casualties and/or damage. Preventing third-party damage.	Yield is 'preparedness' of which value can only be determined implicitly. Baseline alternative not clear. No data on relationship (faster) deployment and effect (rescue and amount of damage prevented).
	Specialist tasks (like scuba diving or hazardous goods incidents)	Visualising return on specific task	• Staff (including pay, training) • Equipment, material • Building • Traffic unsafety prio1 turnout	Deployment in specific accidents. Rescue, damage limitation	Cost difficult to allocate to task, intertwined with basic care. Sensitive due to small number of rescues.

Theme		Possible purpose (S) CBA	Costs	Impact/benefits	Remark
	AFA system (automatic alarms)	Legislative review	Personnel, equipment system and road safety	Assistance and/ or backup during evacuation	Baseline alternative not clear.
	Crisis management	Review or adaptation of legislation	Wide range of costs at various organisations: people and resources	'Trust in government', restoration to situation before crisis.	Baseline alternative missing. Many costs intertwined with basic tasks. Yield not quantifiable.

Table 1.
Application of (S)CBA in fire safety domain.

The Minister of Security and Justice in the Netherlands furthermore recently included two SCBA studies in his consideration and argumentation of choices for fire safety policy measures. These are SCBA studies on policy measures to improve fire safety in residential buildings [14, 15]. The studies examine a number of policy measures to improve fire safety: mandatory residential sprinklers, fire safety requirements for furniture, smoke alarm incentive schemes and fire safety education programs. A cost-benefit balance is presented for the different measures complemented by a sensitivity analysis. For example, both studies show that the cost-benefit balance for residential sprinklers is currently negative mainly because of the (currently) high investment costs. In Akker's study [14], fire safety education programs lead to a positive balance, although it is noted that the (estimated) effectiveness is an important variable. That these studies weigh heavily in the minister's decision is shown by the minister's conclusion in his letter on the subject to the chamber: 'The results of this study do not give the cabinet any reason to proceed with making residential sprinklers compulsory and introducing statutory fire safety requirements for seating and mattresses'.

Both studies are a valuable step in providing insight into the (potential) efficiency of a number of fire safety measures. However, they also show where improvement is still needed. Of the measures studied, quantitative data is needed on their effectiveness and thus yield the following: how many more fatalities or injuries are prevented, and can this be attributed to the measure studied? Both studies recognise this and try to make as educated estimates as possible. In fact, it is often already difficult to demonstrate causality between measure and effect, let alone empirical data on the magnitude of the effect. More frequent use of SCBAs in relation to fire services requires more quantitative empirical data on causality and magnitude of effects. An important effect (or yield) of fire safety measures is the prevention of fatalities. To monetise this effect, both studies use the concept of VOSL. Akker [14] bases the VOSL value on the value of a year of life in good health derived from the health care domain: €2.5 million. Hof [15] uses a VOSL value for road safety adjusted for inflation: €3.3 million. Both studies use VOSL from a different domain and do not further elaborate on the rationale for these choices even though it is an important element in the valuation of revenue.

Finally, both studies address the reduction in fire brigade turnout costs with a reduction in the number of fire brigade turnouts. Akker [14] estimates this item as 'PM' due to lack of data suggesting that it could potentially make a substantial contribution. Hof [15] assumes that these costs decrease proportionally with the number of fires. Here is an important point to note: the costs of the fire brigade's repressive duties are only marginally dependent on the number of fires. Most of the costs are those for preparedness (stations, practice and availability of people and equipment). Only variable costs decrease when there are fewer fires. Section 5 elaborates on some of the above issues.

4. Example case study: Public fire alarm system in the Twente region

To illustrate the application of a social cost-benefit analysis in the fire safety domain, this section describes a case study that is another Dutch example of an SCBA that has been made and used in practice. It concerns an analysis of the costs and benefits of the automatic fire alarm system in the Twente region. This section is based on the article 'Public fire alarm system: history, costs and benefits' [16].

First, the context, origin and purpose of the automatic fire alarm system are briefly described. This is followed by the method used to determine costs and safety benefits. The section concludes with the results and a brief reflection on the results.

As a rule, a cost-benefit analysis mostly deals with a new (safety) measure to be taken and possible alternatives compared to the current situation. The example differs from this in that it concerns an existing safety measure. This means that the historical context surrounding the origin and intended purpose of the safety measure deserves extra attention. This is necessary in order to be able to focus on the 'safety benefits' in particular.

4.1 Introduction AFA

From the beginning of recorded history, fast and accurate reporting of a fire is regarded as essential to control those fires. Based on this notion, automatic fire alarms to public fire departments in the Netherlands originated in the late nineteenth century out of the need to guarantee the alarm to the fire service. During that period of time, an elaborate network of public automatic fire alarms was installed: first in public spaces but shortly afterwards also in various buildings. In the twentieth century with the invention and commissioning of the (wireless) telephone and the increased quality of building fire safety, the relevance for automatic fire alarms in the public domain decreased. Where the fire alarms in public spaces were removed, most of the automatic fire alarms in buildings remained: at first as a public service but later on as an obligation by building regulations.

At the start, the purpose of the automatic fire alarm system (AFA system) was clear and simple: a technical means to automatically guarantee the swift and accurate alarm of firefighters by lack of a reliable alternative. The AFA system is part of the fire detection and alarm systems inside a building. This is illustrated by **Figure 1**.

The benefits of this safety measure for society were obvious: the faster and more reliable the alarm, the better the chance to fight the fire and save lives. However, as alternative means for alarming the fire brigade became available (telephone and mobile phones), the purpose of the (obligatory) automatic fire alarms as described

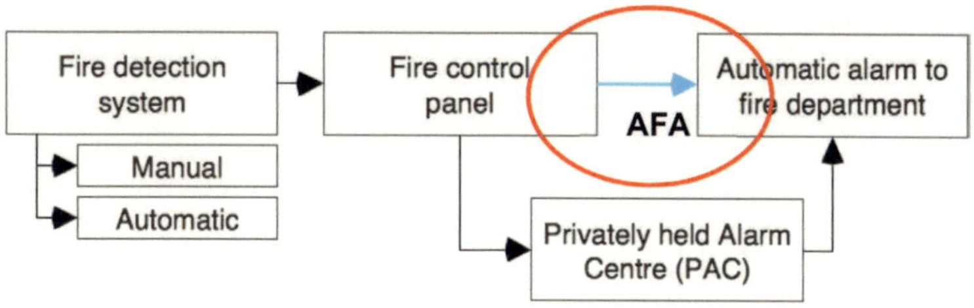

Figure 1.
Schematic overview of the AFA system.

in building regulations became less clear over the years: the purpose changing from 'rapid fire fighting' to 'life-saving' to 'assistance for evacuation'. Recent Dutch building regulations state the purpose for automatic fire alarms to the public fire department as: 'supplementary assistance of fire brigade for rescue and evacuation'.

Another issue concerning AFA systems is that since the beginning of the use of fire alarms, the AFA systems also have caused false and unwanted alarms. Over the years, 2007–2013, 56–62% of all fire alarms proved to be false; 79–87% was caused by an AFA system (see **Table 2**).

Unfortunately, these numbers aren't unique in the Netherlands. Fire departments in the United Kingdom and New Zealand [8, 17, 18] also report such numbers and thereby underline the fierceness of the problem. Inevitably the great number of false alarms causes disadvantages such as demoralisation of fire department staff, the decrease of operational preparedness for other confirmed calls and the increase of risk to the public and firefighters during emergency response.

Where does this take us in relation to costs and benefits? The changing context for AFA systems and the burden of the false alarms raise questions on the cost-effectiveness of this safety measure. Taking a closer look at the intended purpose and costs for objects and society, the question is raised whether the AFA system safety benefits can bear the weight of the costs. A more specific question is can the safety benefits be made visible and do they balance with the imposed costs. To answer this question, a cost-benefit approach can be helpful to gain insight.

This case study uses the cost-benefit approach to assess the added value of the AFA system in the Netherlands. We took the region of Twente as our case study to assess the overall costs as a result of the AFA system and the possible benefits it yields. The reason for this choice is that for this region the authors have access to all required fire

Year	Total fire calls	False or unwanted	False or unwanted from AFA
2007	108,100	60,800 (56%)	52,130 (86%)
2008	109,400	64,100 (59%)	55,645 (87%)
2009	109,600	62,500 (57%)	52,946 (85%)
2010	103,200	61,100 (59%)	48,099 (79%)
2011	107,200	65,200 (61%)	54,093 (83%)
2012	96,000	59,800 (62%)	49,603 (83%)
2013	86,700	50,600 (58%)	41,978 (83%)

Table 2.
False and unwanted fire alarms in the Netherlands (source: www.cbs.nl).

brigade data. To assess the added value of the AFA system, we compare the costs and benefits of the existing situation to a scenario without automatic fire alarms to public fire departments.

The safety region of Twente covers 1632 km^2 and has one fire brigade working from 29 fire stations. On December 31st 2012, a total of 854 buildings (i.e. objects) of various kinds were linked to the regional dispatch centre via an AFA system. Of these 854 objects, only 6% was connected with an AFA voluntarily; the other 94% were mandatory because of the building requirements. The yearly cost related to this connection is approximately 1100 euro per object.

As **Table 3** shows, during the period between 2008 and 2013, the region experienced 17.418 fire reports via an AFA system of which only 560 were marked as 'real fire'. The other 97% of AFA system calls appeared to be false or unwanted. This data comes from the database containing details of each call (including the cause of the AFA alarm) reported by the commanding officer. Classification includes categories such as 'cooking', 'real fire', and 'smoking'.

Even though the extent of false calls as a result of AFA systems is comparable to the figures for the rest of the Netherlands (as shown in **Table 2**), it is hard to determine to what degree the case of Twente as a whole is directly applicable for the rest of the Netherlands. The region of Twente can be characterised as average for the situation in the Netherlands on aspects as densely vs. sparsely populated areas and variation of building types (from small residential to large industrial). To summarise, we believe that the situation with false alarms from AFA systems is representative for the rest of the Netherlands. The results of the cost-benefit analysis therefore give a representative picture for Twente and can be seen as an indication for the Netherlands as a whole.

4.2 Method

Over the years, the high number of false alarms as caused by the AFA systems was given high priority by the Dutch fire departments. As a consequence, many initiatives have been undertaken to decrease this number. Despite this effort, little is known about the actual balance between the costs that are being made to establish and maintain the AFA system and the benefits that it yields: the Netherlands is no exception to this matter. The studies that attempt to establish a cost-benefit analysis acknowledge that this is difficult due to a lack of usable data on safety benefits for such an analysis [8, 18].

In this case study, the costs and benefits of the current situation (the safety measure 'mandatory automatic call forwarding') were compared with the

Number of incidents in the dispatch centre of the Twente fire department Period 1/1/2008-2031/12/2013	
Total number of fire & incident reports	37.507
Number of calls for fires	25.888
Number of calls as a result of the AFA system	17.418
Number of AFA calls that proved false/unwanted	16.858
Number of AFA calls with confirmed fire	560

Table 3.
Incidents in the Twente region 2008-2013.

alternative 'no automatic call forwarding'. The analysis was framed as an economic evaluation of an existing measure, looking to the past for the benefits. Other preventive measures related to fire detection and alarm, such as the fire alarm system, are not part of the analysis. We use the framework of a cost-benefit analysis as introduced by MacDonald [8] in which he discusses alternatives for the automatic fire alarms. In his study, MacDonald describes more specifically what alternatives the fire departments have to the AFA system and how the costs relate to the benefits in comparative perspective. The different categories by MacDonald are to some extent used for the following analysis and, in relation to Lamm et al. [18], formulated as follows:

1. AFA system: costs related to the system itself (e.g. installation, maintenance)

2. Response: costs that are made for each call by the fire department and the risks for society with regards to road safety

3. Results and effects: benefits as a result of the automatic fire alarm

To quantify these three categories, we used our knowledge in the field of fire services while assessing the relevant costs and benefits. Throughout the upcoming paragraph, we will motivate our choices with regard to the quantification of the three categories. For detailed reference, see Ref. [16].

4.2.1 AFA system

The main concern with the AFA system is the costs that are made to maintain the system. As previously mentioned, object owners were obligated to get a subscription to the AFA system and to be directly linked to the public fire department. In the region of Twente, the annual fee amounted to 1100 euro per subscription. On top of that they are required to keep the connection with the central alarm unit online 24/7, which costs approximately 194 euro per year (as paid by the subscriber). Besides this, a so-called TOOM project was launched to decrease the number of false alarms (by instructing subscribers, controlling the objects and other maintenance-related activities). A brief analysis showed that in 2009 alone 3500 man-hours were spend (75 euro per hour, overhead not included) and for the subscriber 8 hours per year (43 euro per hour, no overhead included). When a situation is created in which the AFA system is removed from all objects, the TOOM project can be downsised as well. Based on their experience, the authors use an estimation that 10% of the false alarms will remain.

4.2.2 Response

Under 'response', the costs are enumerated that the fire department is confronted with on every call caused by the AFA system. For the emergency control room alone, the number of extra man-hours caused by alarms via the AFA systems is estimated at 1500 (with an hourly rate of 64 euro). In the period of 2008–2011, a yearly average of 3124 calls were recorded in the Twente region for which, in each case, six firefighters and one truck were used. Per call, we calculated roughly 1 hour and a wage of 25 euro per fire fighter per hour. For the voluntarily staffed fire stations (which in 2008 accounted for 60% of the total number of stations in the Twente region) the same calculation was used. As far as the equipment concerns, only 5 euro was calculated as expense for gas per call. Depreciation of equipment was left out of the calculation since a quantitative relation between the extra call because of the AFA system and additional depreciation of material is hard to quantify (see for instance ([18, 19], p. 21): 'the bulk of the fire service costs are standing army requirements associated with fire risk'.)

The call of a fire truck is accompanied with additional accident risk on the public road. To quantify this, a study was conducted by Rosmuller et al. on the safety of fire truck drivers and the associated traffic-related deaths of citizens [20] and the annual costs as a result of traffic accidents [21]. Based on this, a calculation was made: 3,5 casualties per 179.015 calls and the 12.5 million euro only one casualty costs. Adding up, this makes that, in sum, one call goes for 244 euro. This is a significant but hardly mentioned effect of the AFA systems for society as a whole. The additional costs as result of the risks to the firefighters due to the AFA system are left out of the analysis. Since this risk (most of the time) only occurs in the situation of a real incident, it is assumed that the costs are equal with or without an AFA system.

4.2.3 Safety benefits of the AFA system

Making the 'safety' benefits of the AFA system clear is a complex question, especially when it comes to expressing it in monetary terms. Examples of such an approach are given by previous studies such as Ref. [22]. As later will appear, no lives were saved as a benefit of the AFA system. Therefore, the discussion of how to express the 'saved' live in monetary terms is not elaborated further here. In this case study, the current benefits of the AFA system are made visible based on an analysis of incident reports related to the actions undertaken by the fire department.

To illustrate the benefits of the AFA, all AFA calls classified as a 'real fire' (n = 560) by the fire department within the safety region of Twente in the period of 1/1/2008 and 31/12/2013 were classified based on the actions performed by the fire department. Based on the operational information in the incident report, each call was analysed. A system of four classifications was used based upon the basic tasks of the fire department: varying from actively rescuing of persons and saving the property to a somewhat passive stance in which, due to the limited severity of the fire, no action besides checking general safety and verifying the situation, is undertaken (see **Table 4** for classifications used).

More precise, category 1 implies that the firefighters actually save or evacuate persons and extinguish the fire to prevent further damage to the object based on

	The call led to action by the fire department to prevent expansion of the fire or to limit the damage or to extinguish the fire.	**The call does not lead to action by the fire department to prevent expansion of the fire or to limit the damage or to extinguish the fire.**
The call leads to action by the fire department for rescue and/or evacuation.	1	3
The call does not lead to action by the fire department for rescue and/or evacuation.	2	4

Table 4.
Classifications used in analysing fire reports.

an AFA call. Category two and three are classified when the fire department had to either prevent expansion of the fire to limit the damage by extinguishing (2) or save and evacuate persons (3). The fourth category means that no action was necessary to either put out the fire or that the firefighters had to save or evacuate persons. In day-to-day operations, the responsibility of the fire department for calls in category 4 lies with securing, ventilating and checking the premises and making sure everything is safe. In these cases, often the fire has been extinguished by the people in the building themselves (e.g. internal safety staff).

In case of doubt on the information in the incident report on which party extinguished the fire (fire department or internal safety staff), the call was classified as type 2: fire department. In other words, the fire department gets the benefit of the doubt to ensure no underestimation of fire department actions was created.

4.3 Results

4.3.1 Costs of AFA system and response

Table 5 shows a summary of the estimated costs for the current situation of the AFA system and response. The rightmost column shows the situation in which no AFA system is put into practice with the assumption that 10% of the false alarms remains[1]. This same table reveals that the AFA system in the region of Twente costs' society approximately 3 million euro per year. It is important to note that this is an estimate based on a conservative approach. Stopping with AFA systems leads to lower costs. Some costs will fall to zero (AFA system technique, monitoring connections) and some will remain or will be reduced (effort on false alarms, dispatch centre and response). We acknowledge that even in a situation without the AFA system there will be a number of false alarms. Even so, the total number of expenditure will be lower as showed by **Table 5**. This is at least the direct result of saving on the maintenance of the AFA network and all the technique involved.

4.3.2 Safety benefits as a result of the AFA

To determine the AFA safety benefits, a total of 560 incident reports from the safety region of Twente were reviewed for this case study. Each incident was classified into a category as summarised in **Table 4**. **Figure 2** gives an overview of the results:

From these 560 incidents, 14 reports did not contain adequate information to determine a classification. The remaining 546 reports were successfully classified 1 to 4. This leads to 4 incidents in category 1, meaning that in these incidents the fire department had a task in rescue or evacuation. In 3 cases, the fire brigade assisted in and verified the evacuation. In 1 situation, a missing person was rescued. About 128 incidents were categorised as '2' (only firefighting) and the majority of 414 were assigned to category '4', which implies that no action was undertaken by the firefighters other than verification or ventilation activities.

The AFA alarms that are reported always receive immediate priority meaning that in every call a fire truck is sent to the object. Subsequently, the object is

---
[1] This assumption of false alarms is based on the authors' experience on the part of false alarms via telephone. Additional research is necessary.

	Description	Current AFA situation (3124 calls/yr)	No AFA system (10% of false alarms remain)
AFA system	Technique and administration	€ 607.035	—
	Costs dispatch centre	€ 326.865	€ 326.865
	Connection monitoring	€ 164.706	—
	TOOM fire department	€ 262.500	€ 26.250
	TOOM object owners	€ 283.112	€ 28.294
		€ 1.644.218	€ 381.409
Response	Dispatch centre	€ 96.000	€ 9.600
	Response firefighters	€ 765.380	€ 101.675
	Traffic accidents	€ 762.275	€ 101.263
		€ 1.623.655	€ 212.538
Total		€ 3.267.873	€ 593.947

Table 5.
Summary of costs related to the AFA system.

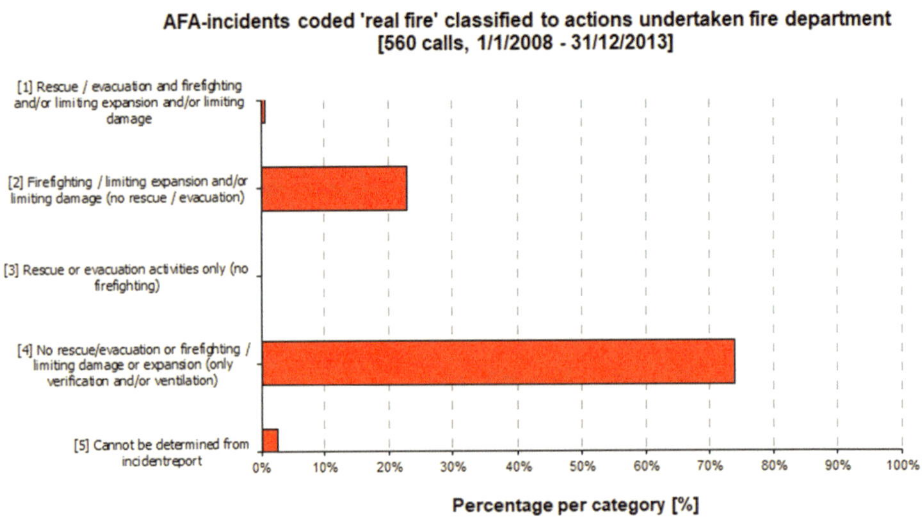

Figure 2.
AFA incident classified to actions of the fire department.

contacted to get information concerning the severity of the call. In 2008 alone, 53% of the calls were deprioritised while the firefighters were making their way to the object. In 23% of the 560 cases, the fire department had to put out a fire: in 84% of these, the commanding officer reported that fire was small and could be handled by one unit (meaning that no additional fire trucks needed to be alarmed). The majority of the reported fires was limited to an item or object (e.g. like a magnetron, oven or ashtray).

When we look at these results in relation to the goal of the AFA system, it appears that in a 6-year period only 4 incidents occurred in which the fire department had to take action in rescue or evacuation. In the same period, the firefighters only extinguished 128 emerging fires: it is hard to argue that these premature fires could—or could not—have turned out to become severe without the AFA system. Therefore we acknowledge that, with the existing database, we are only able to talk about the current situation in which AFA does exist and not a situation in which it does not exist.

Based on the case study, we estimate that on a yearly basis the AFA system in the Twente region costs 3 million euro. If compared to safety benefits related to the goal of the AFA system, 'providing supplementary assistance with the evacuation', these costs do not outweigh the benefits. Even so, the question stands whether a faster appearance by the fire department increases the number of rescues and therefore overall safety. Unfortunately, the correct and usable data to support such a statement is missing in the Netherlands and therefore no valid statement can be made. If 'monetary break even' is taken as the goal the estimated costs represent the minimum increase in fire safety (or safety benefits) the AFA-system should at least 'produce' in term of money. Suurmond and van Velthoven [22] used this 'break even' approach in their study to the increase in safety in the hospitality industry. In our case, the positive effects in terms of safety in the region of Twente at least have to break 3 million in order to be cost-neutral. Based on this finding and the limited safety benefits, the recommendation will hold ground that *at least* the policy with regard to a mandatory AFA system for most buildings has to be analysed and adjusted. The central goal of this adjustment should, as far as the authors are concerned, put the responsibility with the objects' owner and not the government or fire department. Again, in only 23% of the cases classified as 'real fire', the firefighters had to put out an emerging fire.

The overview of costs shows that there is potential for cost savings in the area of traffic safety. Limiting unnecessary firefighter turnouts leads to a safety benefit. In 25% of cases, the fire brigade extinguished an incipient fire. While the available data does not allow an estimate of the potential economic benefits, this can be qualified as a safety benefit of the AFA system.

A sharp quantitative goal for the AFA system was not found. However, it does appear that the qualitative definition of the objective has changed over time. The question is what this should mean in practice for the fire brigade. The current wording places the responsibility primarily on the user and at the very least offers room for the fire brigade to take a critical look at its own role. Contrasted with the research findings, a basic response to an AFA notification 'go immediately to the scene with horns and bells for rescue' is too pretentious.

Mandatory forwarding was and is intended to speed up the alerting of the fire brigade. 'Faster' here is a relative term. Relative in relation to the steady growth in the availability of phones and the state of the art. In addition, it is relative because the speed of the fire brigade is influenced by many more factors. Consider, for instance, the time taken at the control room, the time it takes volunteers to get to the station and the time it takes to drive to the fire address. Also, as a result of the development in the field of in-house emergency response, institutions and companies are prepared for rapid notification of calamities and their initial combating. It has been found that the internal organisation at the company or institution often already has the situation under control. This immediately shows the added value of the FDS and associated internal alarm.

4.4 Use of the SCBA in the political decision-making

The basic findings of this case study on the safety benefits of AFA systems were presented in 2010 in Netherlands [16]. Based on the absence of 'safety benefits' and a strong negative cost-benefit balance, it contained the recommendation to policy makers to critically review the existing building regulations regarding AFA systems. By removing the AFA system from the building regulations, society would benefit from the cut down in costs for maintaining the system without seriously compromising fire safety.

In the period between 2009 and 2012, the Dutch minister of Interior Affairs was working on a broader package of changing the building regulations. This was as part of a Government-wide process to decrease the number of regulations. The regulations concerning AFA systems were part of this package. The minister of Interior affairs proposed in 2009 to strongly reduce the number of objects requiring an AFA connection. In 2012, the Dutch building regulations had changed. The findings of the cost-benefit approach were used as a reference in the relevant documents (i.e. regulation itself and the supplementary documents) describing and motivating the adjusted regulations.

The observation that an AFA system does not demonstrably improve safety and does cost a significant amount of money led to an amendment that removed the obligation for certain buildings to subscribe to an AFA service of any kind. This directly implied that some of the responsibility was shifted from the fire department to the object owners.

Although from the SCBA case it can be concluded that all AFA connections can be removed, the political decision was to keep the AFA connections in place for certain objects seen as more vulnerable (like child day care centres). This shows that the outcome of a CBA is just one of the inputs for the decision on a policy change involved. Additional arguments in this case that are difficult to value in a CBA are commercial interest of companies selling and installing AFA systems, the safety perception that comes from having a direct line to the fire service and the fire service feeling itself responsible as a 'back-up safety for society' just in case.

Nevertheless, the SCBA of the AFA system has, in line with Sunstein [13], 'turned on the light' in this specific case. It has now become clear that although the countable safety benefits are very small, we still choose to keep the AFA regulations in place for some specific buildings. This has been an explicit consideration where in this case the preferred option is 'second best' from an economic perspective. As a result of the adjusted building regulations, the number of AFA connections in the region of Twente dropped in the period between 2013 and 2015 from approx. 830 to roughly 500. On a national scale, the effect of the changed building regulations has become visible by a reduction in AFA calls.

5. Some relevant SCBA themes considered in more detail

The previous sections discussed the application of the social cost-benefit analysis tool within the firefighting domain and presented an elaborate case study of SCBA. The following section elaborates on some specific themes in relation to using and preparing an SCBA within the firefighting domain.

5.1 Safety yield and null alternative

The effects of public policies are often difficult to measure. De Bruijn [23] cites four reasons for this: public tasks often have to serve multiple values, the effect arises in co-production with third parties, the period between the intervention and the result can be long and more abstract goals are often involved. This also applies to effects of firefighting. Even objective safety (countable effects such as fewer casualties and less damage) involves effects that often only become visible in the long term and can also be influenced by other factors. They may also be effects that do not become visible at all. In fire safety, for example, 'preventing' is an important but difficult to make visible effect. In addition, fire safety contributes to 'sense of safety' and tries to contribute to 'safety awareness'. These are effects that are abstract, subjective and difficult to measure.

Making the safety benefit of firefighting or fire safety measures concrete is therefore not easy. See for example a slightly dated but still relevant article of Challands into effects of faster response times and fire outcomes [24]. He concludes that research on this topic is limited. After extensive research into 27,500 building fires in New Zealand during 2003–2008, he cannot prove a relation between a higher response time for building fires and more (deadly) casualties. In addition, research from Saylors [25] is noteworthy. In a study into the added economic value of a local fire brigade, Saylors describes a model in which he tries to quantify the number of buildings that would have burned down if there had been no fire service intervention. Saylors takes into account the buildings that have been saved by the fire service intervention ('nearby at-risk properties'). He introduces the 'saved ratio': the relation between the (economic) value of the saved buildings divided by the value of the total number of buildings 'at risk'. The 'saved ratio' was calculated for four cases. His conclusion is that the return on investment of the local fire brigade is 2200% (the costs for the local fire service are a factor 2200 lower than the yearly prevented potential damage). Saylors says: 'to change the dialogue of public safety by recognising that the worth of a fire department is greater than a calculation of loss and that the worth of a fire department is more accurately represented by including what is saved' ([25], p. 46).

To make cost-benefit analyses, it is important to have research into the effects of fire safety measures. In doing so, it is in any case essential to have registrations of 'output' within the fire service field in good order. The investigation in the case study showed, among other things, that the registrations of fire service incidents in the joint control room system are useful for this purpose. With this GMS database, the fire service has a large amount of practice data at its disposal. To increase the usability for further research, it is important to control the content, uniformity, correctness and completeness of the registrations in advance.

Although there is (limited) data on the number of fire victims, visibility into the null alternative, a key reference point in an SCBA, is limited. Especially for assessing existing policies, this is a major limitation. What safety benefits does the fire service realise with the current policy? On top of that, it is not easy to determine how much damage is prevented by the fire brigade in the current situation or how many casualties are avoided.

5.2 Being clear about history and purpose of fire safety measures

To determine the yield or effect in an SCBA, it is necessary to have the objective of a fire safety measure as clear as possible. In the fire service's field of work, it is often not exactly clear for each task what the objective has been historically, let alone that the trade-off (in costs and benefits) is clear. If an SCBA is then carried out in the here and now for the diving task, for example, it is advisable to consider the history surrounding the creation of that task. In that respect, the benefits or effects cannot always be quantified and monetised.

5.3 The value of standing army requirements

For the fire service, preparedness, 'being there', and effort, 'someone is still trying their hardest', are also a societal benefit. Its explicit value is difficult to determine, at most implicitly via the costs to be accepted. Many costs within the fire service are therefore 'fixed' and not dependent on number of deployments. In that respect, firefighting is a kind of 'collective insurance'. Related to this, using cost per turnout in SCBA by dividing a fire service budget on the number of turnouts therefore makes little sense. With no turnouts, the costs are not zero. If this element is not properly taken into account, it does not provide a useful SCBA. Also, the fire brigade combines many tasks, which makes isolating costs per task difficult or gives an incorrect picture (e.g. scuba diving task, staffing of crisis management teams). Stopping or reducing a particular task does not always result in a proportionate release of resources.

5.4 Fire safety and VOSL

An essential part of a social cost-benefit analysis is the translation of all inventoried costs and benefits into a monetary unit. Through this translation, the balance of costs and benefits can be calculated. For some topics, this is not a problem: the costs of a fire station or volunteer allowances can easily be expressed in monetary terms. However, there are also costs and benefits where this is less straightforward, especially when it comes to fire safety. For example, the 'cost' of traffic unsafety casualties caused by fire brigades' prio 1 calls or the 'revenue' of rescued casualties or prevented injuries. For both traffic unsafety and rescued casualties, it is necessary to assign a monetary value.

Economists have developed the 'value of a statistical life' for quantifying and standardising this. The abbreviation used for this is VOSL: value of a statistical life [26–29]. The term 'statistical' is meant to make it clear that this is a technical term. The concept of VOSL is (of course) not about the value of an identifiable human life but, as the term indicates, about a 'statistical life'. The crux is essentially the amount of money deemed by an individual or society to be in proportion to a minor adjustment to a safety or health risk: the probability of death, injury or illness. It is the amount of money that people are willing to pay for a given reduction in accident risk (the so-called 'willingness to pay' [WTP]).

VOSL is then calculated by dividing the WTP by the risk reduction (Δp): VOSL = WTP/Δp. A literature review by the OECD on the basis of willingness to pay studies[2] recommends for EU countries a VOSL value of US$ (2005) 3.6 million (range: 1.8–5.4)[3] [30].

Assigning a value to a human life does encounter resistance on the principle. From the perspective of economists, it is not strange: without knowing it, we constantly assign a price to a human life when taking risks ourselves on a daily basis. It would go too far to go into that aspect of VOSL in detail in this chapter. Clearly, a monetary valuation is necessary to make considerations in risk policy or more specifically within firefighting. Often, such a value is implicitly part of the decision-making process. By deciding whether or not to invest in a new fire station, the costs (building, vehicles and personnel) are weighed up against the fire risk to be covered (number of turnouts and possible prevention of casualties and/or damage). In doing so, the competent administrative body implicitly assigns a value to human life.

---

[2] The basis for the study comprises a raw database of 856 VOSL estimates from 1970 onward.

[3] For application of this value, inflation correction is required, among other things ([30], p. 127).

In contrast to for example road safety [1, 31, 32], it is surprising to note that little research has been carried out into VOSL values for fire safety. A study commissioned by the New Zealand Fire Service provides a preliminary summary of VOSL values within the international fire service domain: '*The international literature contains a broad range of VOSL values across different risks; different measurement techniques; and for different populations. Unfortunately, there are few estimates of fire VOSLs*' [33]. At the same time, Sanderson et al. [33] conclude that the existing studies do not estimate a 'stand-alone' value for fire safety but instead translate the VOSL values from other studies (such as road safety) into a fire context. A more recent study commissioned by the European Commision to a broad consortium of research groups also calls for more costs-benefit analysis (CBA) to base fire safety measures upon [34]. The study, with the main focus as the improvement of fire statistics on a European scale, analyses three case studies using CBA for fire safety measures. It concludes that rightfully so far road safety VOSLs were used in fire safety CBAs because they have been studied much more and thus are considered more reliable. However, the need for a fire safety VOSL value is also stipulated.

Taking this as a starting point, De Wit et al. summarised and updated the literature on fire safety VOSL [35]. The results in **Table 6** are taken from this article:

Author(s)	Country	Method and context	Estimation $VOSL_{fire}$
Dardis [36]	United States	• Revealed preference • Consumer market smoke alarms • House fire	US$ 0.1–0.68 million (1976)
Garbacz [37]	United States	• Revealed preference • Consumer market smoke alarms • House fire	US$ 2.7 million (1986)
Chilton et al. [38]	Great Britain	• Stated preference • Contingent valuation • Various, including house fire and fire in public area	*1998 study:* House fire: 0.88–0.93 x $VOSL_{road}$ Fire public area: 0.92–0.92 x $VOSL_{road}$ *2000 follow-up study:* *House fire: 0.84*–0.89 x $VOSL_{road}$ Fire public area: 0.91–0.96 x $VOSL_{road}$
Sanderson et al. [33]	New Zealand	• Stated preference • Literature review • House fire	Relative: 0.57–0.66 x $VOSL_{road}$
Carlsson et al. [39]	Sweden	• Stated preference • Choice experiment • Various, including house fire	EUR 1.4 million (2010) Relative: 0.66 x $VOSL_{road}$

Table 6.
Literature overview – VOSL fire safety.

In addition, De Wit et al. conducted a choice experiment among Dutch citizens to estimate a 'Fire VOSL'. Similar to earlier VOSL research of De Blaeij and Carlsson et al. [31, 39], the study uses the stated preference approach to estimate willingness to pay for risk reductions by means of a choice experiment via internet survey among the Dutch population. The objective of the study was to arrive at a VOSL estimation in relation to house fires. In the choice experiment, respondents had to make a series of choices between the attributes 'risk of death from fire' and 'increased taxation for safety measures'. A questionnaire was designed and distributed among approximately 500 Dutch citizens of 18 years and older. The data from the choice experiment contains choices made by a total of 504 respondents who each made 9 discrete choices. This means that on the basis of these choices in which the respondents showed their preferences for changes to their risk and paying more tax, utility parameters could be estimated from which VOSL can subsequently be derived. For a detailed description of theory and method, reference is made to [35]. The results show that median VOSL ranges from 8.9 to 15 million euro, while average VOSL ranges from 12.2 to 21.9 million euro. The estimate of these VOSLs is based on economic valuation using a stated preference technique. Authors opted for a choice experiment in which respondents were asked to choose between various hypothetical scenarios. The choice experiment provided various estimates for VOSL in regard to house fires. The mean of the sample in which respondents' protest answers were removed produced a VOSL of 43 million euro. Protest answers refer to respondents who opted for the present situation in all their answers and who gave as a reason, for example, that they objected to paying more tax in any situation. It also transpired that some respondents in the choice experiment in all nine questions opted for 'safety improvement' regardless of a tax increase. This choice behaviour seems to resemble what the literature describes as 'payment vehicle non-attendance'. This choice behaviour can reflect genuine preferences but can also result from being presented with a hypothetical situation in which payment is not actually required (so-called hypothetical bias). This clearly is an issue for further study. Removal of these respondents from the sample results in a considerable reduction of VOSL; VOSL is then 22 million euro.

The present study comprises the first-choice experiment in the Netherlands with 'probability of dying' in regard to this subject. The results of this exploratory study are that people seem to be willing to pay an order of magnitude more for preventing death by fire than by road accidents. This is fitting in with the observation of De Blaeij [31] that VOSLs differ between types of risks. It however quite contrasts with Carlsson et al. [39] who found that VOSL of fire was one-third of that of road safety. The experiment thus needs to be repeated and updated in order to better assess the reliability of the outcome and should include attention to the wording and presentation of the hypothetical scenario.

6. Conclusion

To explicitly identify the costs and benefits of safety policy, the social cost-benefit analysis tool can also be put to good use within the fire safety domain. The examples of this are now available and used to support political-administrative choices on fire safety policy.

When making and applying social cost-benefit analyses to firefighting, a number of specific themes require extra attention. Some costs are not directly visible at first glance but may have a significant share in the cost-benefit balance. For example, the case study shows that the social costs resulting from traffic unsafety due to high priority fire brigade turnouts can be significant relative to other costs. The same example also brings to light a further point: reliable quantitative data on effects is needed to capture costs and benefits. Registrations of traffic accidents caused by fire brigade turnouts, for example, have only recently become the subject of registrations again.

An important goal of firefighting is to prevent accidents. In the context of SCBA, this is a tricky issue. How can the benefit of prevented damage or casualties be quantified?

Fire service costs are largely made up of (fixed) costs for preparedness: being available to turn out. In that respect, firefighting is a kind of 'collective insurance'. Stopping or reducing a particular task then does not always yield proportionate releases of resources. If this element is not properly taken into account, it distorts the results of an SCBA.

VOSL plays an important role in monetising one of the most important 'returns' of fire service work: the prevention of (fatal) casualties. From that perspective, it is important to have a substantiated VOSL value for use in cost-benefit analyses for firefighting.

Given the limited research on 'fire VOSL' when using the VOSL approach in the decision-making process for a fire safety policy, we side with the consortium of researchers that concluded that VOSL for road safety is best researched and therefore should be used as a basis for cost-benefit analysis for fire safety [30] However, we additionally advise to both present the outcome of calculations using the road safety VOSL value and the derived fire safety VOSLs from the literature. This can be a stand-alone fire VOSL from Refs. [35] or [39] or the relative value related to road safety from Ref. [33]. Wider application of SCBAs within fire services requires further research into VOSL values for 'fire'.

The SCBA is widely used in other places, an advantage of this is that a lot of information is available on the basic methodology for making an SCBA. In the previous chapter, a number of references were provided for this purpose.

Conflict of interest

The authors declare no conflict of interest.

Author details

Ron A.C. de Wit[1*] and Ira Helsloot[2,3,4]

1 Veiligheidsregio Twente, Enschede, The Netherlands

2 Radboud Universiteit, Nijmegen, The Netherlands

3 IMT Mines, Albi, France

4 Technische Universität Darmstadt, Germany

*Address all correspondence to: r.dewit@brandweertwente.nl

References

[1] Elvik R. Cost-benefit analysis of road safety measures: Applicability and controversies. Accident Analysis & Prevention. 2001;**33**(1):9-17

[2] Van Velthoven BCJ. Cost-benefit analysis of crime policy. Journal of Criminal Justice (Trial). 2008;**4**(87):108-120

[3] McIntosh E, Clarke PM, Frew EJ, Louvière JJ, editors. Applied Methods of Cost-Benefit Analysis in Health Care. Oxford: Oxford University Press; 2010

[4] Van Coile R, Lucherini A, Kumar Chaudary R, Ni S, Unobe D, Gernay T. Cost-benefit analysis in fire safety: State-of-the-art and reference methodology. Safety Science. 2023;**168**:106326

[5] Hewitt M, Biermann F, Greatbach I. The Economic and Social Value of UK Fire and Rescue Services. London: NFCC; 2022

[6] Gernay T, Ni S, Unobe D, et al. Cost-benefit analysis of fire protection in buildings: Application of a present net value approach. Fire Technology. 2023;**59**:2023-2053

[7] McNamee MS, Andersson P. Application of a cost–benefit analysis model to the use of flame retardants. Fire Technology. 2015;**51**:67-83

[8] MacDonald M. Costs and Benefits of Alternative Responses to Automatic Fire Alarms (Fire Research Series 2/2008). London: Department for Communities and Local Government; 2008

[9] Romijn G, Renes G. General Guidance on Social Cost-Benefit Analysis. The Hague: CPB/CPL; 2013

[10] Quah E, Toh R. Cost-Benefit Analysis, Cases and Materials. New York: Routledge; 2012

[11] Ramachandran G. The Economics of Fire Protection. 1st ed. London: Routledge; 1998

[12] Helsloot I. Beyond Symbolism. The Hague: Boom Juridische Uitgevers; 2007

[13] Sunstein CR. Risk and Reason: Safety, Law, and the Environment. New York: Cambridge University Press; 2002

[14] Akker I, Tieben LAW, Bos JGH, Van der Veen MJ. Investing in Fire Safety: SEO Report 2010-78. Amsterdam: SEO Economic Research; 2010

[15] Hof B, Rougoor W, Tieben B. Social Cost-Benefit Analysis of Residential Fire Safety, SEO Report no. 2014-16. Amsterdam: SEO Economic Research; 2014

[16] De Wit RAC, Helsloot I. Public fire alarm system: History, costs and revenue. Tijdschrift voor Veiligheid. 2010;**9**(4):30-45

[17] Johnson D. Operational Response to Automatic Fire Alarms and Amendments to Service Pre-Determined Attendances. Rivenhall: Essex County Fire & Rescue Service. EFA/105/10; 2010

[18] Lamm F, Van Rossum F, Lomax D. The Cost Factors and Profile of False & Unwanted Fire Alarm Activations in New Zealand. Wellington: New Zealand Fire Service; 2006

[19] Goodchild M, Sanderson K, Leung-Wai J, Nana G. The Cost of Managing the Risk of Fire in New Zealand. Wellington:

New Zealand Fire Service Commission; 2005

[20] Rosmuller N, Stoop J, Scholtens A, Elias I, Martens M. Traffic Safety of Firefighters. Arnhem: Nederlands Instituut voor Brandweer en Rampenbestrijding; 2001

[21] SWOV. Factsheet: Costs of Traffic Accidents. Leidschendam: Foundation for Scientific Traffic Safety Research; 2009

[22] Suurmond G, Van Velthoven BCJ. Fire safety in the hospitality industry: Application of cost-benefit analysis in the justice field. Justitiële Verkenningen. 2003;**29**(9):8-29

[23] De Bruijn H. Managing Performance in the Public Sector. 2nd ed. London: Routledge; 2007

[24] Challands N. The relationship between fire service response time and fire outcomes. Fire Technology. 2010;**46**(3):665-676

[25] Saylors E. Quantifying a Negative: How Homeland Security Adds Value. Monterey, California: Naval Postgraduate School; 2015

[26] Ashenfelter O. Measuring the value of a statistical life: Problems and prospects. The Economic Journal. 2006;**116**(510):10-23

[27] Viscusi W, Aldy J. The value of a statistical life: A critical review of market estimates throughout the world. The Journal of Risk and Uncertainty. 2003;**27**(1):5-76

[28] Hultkrantz L, Svensson M. The value of a statistical life in Zweden: Review of the empirical literature. Health Policy. 2012;**108**:302-310

[29] Lindhjem H, Navrud S, Braathen N, Biausque V. Valuing mortality risk reductions from environmental, transport, and health policies: A global meta-analysis of stated preference studies. Risk Analysis. 2011;**31**(9):1381-1407

[30] OECD. Mortality Risk Valuation in Environmental, Health and Transport Policies. Paris: OECD Publishing; 2012

[31] De Blaeij A. The value of a statistical life in road safety [PhD thesis]. Amsterdam: Vrije Universiteit; 2003

[32] Bahamonde-Birke F, Kunert U, Link H. The value of a statistical life in a road safety context—A review of the current literature. Transport Reviews. 2015;**35**(4):488-511

[33] Sanderson K, Goodchild M, Nana G, Slack A. The Value of Statistical Life for Fire Regulatory Impact Statements (Research Report no. 79). Wellington: The New Zealand Fire Commission; 2007

[34] European Commission. EU-Firestat Project Task 6: Case Studies Using Cost/Benefit Assessment Methodology. Vol. 11. Luxembourg: Publication office of the European Union; 2022

[35] De Wit RAC, Helsloot I, Koetse M. The value of a statistical life in reducing fire risk: A choice experiment among dutch citizens. Safety Science. 2024;**16**:106322

[36] Dardis R. The value of a life: New evidence from the marketplace. The American Economic Review. 1980;**70**(5):1077-1082

[37] Garbacz C. Smoke detector effectiveness and the value of saving a life. Economics Letters. 1989;**31**:281-286

[38] Chilton S, Covey J, Hopkins L, Jones-Lee M, Loomes G, Spencer A, et al. Public perceptions of risk and preference-based values of safety. The Journal of Risk and Uncertainty. 2002;**25**(3):211-232

[39] Carlsson F, Daruvala D, Jaldell H. Value of statistical life and cause of accident: A choice experiment. Risk Analysis. 2010;**30**(6):975-986

Chapter 2

Perspective Chapter: Mobile Radio Emergency Communications for Large-Scale Wildfire Fighting – Portugal as a Case Study

Carlos Salema and Rafael Caldeirinha

Abstract

This perspective chapter explores the critical role of voice push-to-talk terrestrial mobile radio emergency communications in supporting firefighters in disaster relief operations. We discuss operational context and challenges, innovative solutions, and future trends in communication technologies for enhancing operational effectiveness and public safety. Each section provides insights and practical recommendations for improving emergency communication systems in firefighting and disaster relief operations, particularly large-scale wildfire fighting, using Portugal as a case study. The chapter comprises five main sections, in addition to Introduction and Conclusions; Operation Context and Challenges; Mission-Critical Communication Networks and Deployment; Network Resilience and Optimisation; Management, Operation, and Security of the Communication Network; and Future Trends in Critical Communications. Mention will be given to deployment strategies and real-world applications of land-mobile radio in emergency communications, including propagation studies on the effects of wildfires in radio communications recently conducted by the authors.

Keywords: mission critical emergency communications, wildfires, land-based radio communications, redundancy, resilience

1. Introduction

Effective communication is crucial for emergency response operations, particularly large-scale wildfire fighting. This chapter explores the vital role of voice push-to-talk terrestrial mobile radio emergency communications in supporting firefighters during disaster relief operations [1]. It addresses the operational context and challenges in deploying and maintaining communication networks in remote and rugged terrains, where traditional communication methods often fall short. Additionally, the chapter delves into the intricacies of mission-critical communication networks, highlighting the importance of network resilience and optimisation to ensure continuous and reliable communication under extreme conditions. By examining real-world applications, case studies, and future trends in communication technologies, this chapter provides comprehensive insights and practical recommendations for enhancing emergency communication systems. The goal is to improve operational effectiveness and public safety, particularly in large-scale wildfire fighting where timely and reliable communication can make the difference between success and failure in managing emergencies. In principle, the national security and emergency network can assume three configurations: private network, public network, and public-private mixed network.

Experience shows that, in extreme situations when national security and emergency networks are most needed, private networks are so congested that they cannot guarantee the desired reliability [1–3]. Additionally, public networks aim to meet the communication needs of the general public, so in regions with low population density, they have a lower degree of coverage, and it is often in these regions where situations (such as fires) that require the most communications occur. For the reasons stated, it is impossible to implement a national security and emergency network exclusively over public networks with current technologies.

For portable terminals and equipment available on the market, it is only technically possible to ensure the desired coverage and availability with a radio network, operating at the lowest possible frequency (in the hundreds of MHz). A network of base stations guarantees spatial coverage. Experience shows that, for the 400 MHz bands, more than 500 base stations are needed to achieve the desired coverage for a country the size and the ground ruggedness of Portugal [1]. In contrast, for the 700 MHz band, the number of base stations required for the same degree of coverage increases, potentially exceeding 700 base stations. This difference arises because higher frequency bands, such as 700 MHz, typically offer less coverage per base station compared to lower frequency bands like 400 MHz; therefore, a greater number of base stations are required to ensure comprehensive and reliable communication across the entire country.

For the network to function, the base stations must be connected to the main nodes of the network where the switches (equipment that allows the interconnection of communications between mobile terminals served by different base stations) are located. This network, for simplicity, will be called the transmission network or backhauling, and will, in principle, be private. Still, it is possible to conceive it, on a case-by-case basis, where sharing with secure high-speed public networks is possible but also safe and economical [4].

The desired coverage guarantee is difficult to achieve only with fixed base stations given the orography of the national territory and the occurrence of extreme events where traffic reaches very high values. Mobile stations are required to respond to these situations and catastrophic failures. These, capable of handling the traffic of any base station, should be pre-positioned across the country according to the risk level forecast, to minimise the travel time to two hours for potential occurrences. A satellite link should ensure the connection of mobile stations to the switches (essential for them to act as base stations).

At present, no standardised technology exists for security and emergency networks that significantly differs from the TETRA technology currently used in most European countries. Adopting a different technology would make the national safety and emergency network dependent on a single supplier [5, 6].

2. Operational context and challenges

2.1 Structural framework and functional dynamics of public protection and disaster relief networks

Internationally, national emergency communication networks are often referred to as Public Protection and Disaster Relief (PPDR) networks. These networks are essential for emergency communications among authorities, especially those critical for their mission fulfilment. PPDR communications, specifically radio communications, referred herein as Mission Critical Push-To-Talk (MCPTT) communications, are divided into two main areas of emergency response activity:

1. *Public protection* (PP): Radio communications used by authorities and organisations responsible for maintaining law and order, protecting life and property, and handling emergencies;

2. *Disaster recovery* (DR): Radio communications used by authorities and organisations tasked with addressing severe societal disruptions that pose significant threats to human life, property, and the environment, whether caused by accidents, natural causes, or human activities, and that may arise suddenly or as a result of a prolonged complex process.

Other emergency communications are managed in a way that, while distinct, needs to be carefully considered and implemented in PPDR systems to ensure overall coherence and the necessary operational coordination. For example:

- *Emergency communications from citizens to authorities* should be handled through calls to 112, the European emergency number;

- *Emergency communications from authorities to citizens* should be managed through public warning systems;

- *Emergency communications among citizens* should be facilitated through public networks and publicly accessible electronic communication services, such as national roaming.

Although authority communications are not limited to systems like PPDR and involve a mix of solutions including public networks or other private state networks, the expectation is that PPDR systems should serve as the guarantor of emergency and security communications, ensuring that its user authorities can rely on it for fulfilling their missions and operations regardless of the operational context.

In developing this chapter, a set of key concepts was considered important to clarify, as follows:

1. *Management and operation of PPDR systems* encompass all actions and processes necessary to ensure their continuous proper function, complying with user operational and functional requirements. The security of PPDR systems is defined by their ability to withstand actions that could compromise the availability, authenticity, integrity, or confidentiality of the system, its data, or its services.

2. *A security incident* is identified as any event that could adversely affect the security of PPDR systems. The assets of these systems include management, operation, and security centers, radio stations, switches, transmission (backhauling) or information systems, equipment, infrastructure, software applications, and all physical, logical, and human resources supporting the system, including interconnections and third-party contracts.

3. *A Single Point of Failure* is an asset without redundancy whose failure, especially due to a security incident, causes a significant adverse impact on functionality, including potential interruptions or security compromises. Preparation involves activities, contingencies, and measures taken before an incident/disaster to ensure an effective response to its impacts. Resilience is the ability of an organisation, resource, or structure to withstand internal and external threats, endure partial capacity loss, and recover service delivery with minimal reasonable performance loss. Resilience aims to maximise the mean time between failures (MTBF) and minimise the mean time to repair (MTTR), applying across all system levels, including components, systems, and subsystems such as switching, transmission, and terminal equipment. An example of a severe backhaul failure during a wildfire is illustrated in **Figure 1**. The wooden pole supporting the optical cable burned at its base, leading to the collapse of the entire setup and resulting in a disruption of the backhaul link. Consequently, the connecting base station switched to local site trunking (LST) mode.

4. *Redundancy* refers to the additional resources beyond what is necessary to ensure that a particular functional element performs its required function or data set preserves information. The characteristics of PPDR systems, particularly man-

Figure 1.
Single point of failure: Optical fibre backhaul link affected during a wildfire.

agement, operation, and security, are shaped by the operational context of the user entities and the actions they must undertake to fulfil their missions. These actions often involve more than one entity, may be joint operations, and can occur under extraordinary circumstances resulting from extreme events or human actions, including intentional acts.

Accordingly, PPDR systems must be prepared and equipped with the means and capabilities to guarantee, at any given moment under varying operational conditions, the necessary emergency and security communication functions and services for ongoing and future operations, especially highlighting command and control of forces, understanding of the operational situation, and upward information exchange in the hierarchical line and among forces involved in an operation within the intervention zone.

2.2 Entities and scope of operation

Public safety and emergency network users are typically defined within national legislation or governmental resolutions. However, it is crucial to acknowledge and incorporate developments occurring since the inception of these documents. This includes integrating entities that, while not current users, play significant roles within the operational context as outlined by the relevant legal framework. Legislation related to civil protection, internal security, emergency planning, and national defence is critical in ensuring communication support for connections to national, regional, and local level, governmental structures.

PPDR networks across European countries should include a wide range of entities, both public and private, involved at various governmental levels. These entities, responsible for decision-making and operational execution in civil protection, internal security, and emergency planning, must have reliable access to emergency and security communications in the effective fulfilment of their roles. Moreover, PPDR systems are tasked with ensuring robust connectivity for all listed entities to essential decision-making facilities and resources. This comprehensive connectivity spans command centers, headquarters, operations, and communications rooms. It extends to maritime, terrestrial, and aerial assets employed in actions or operations across the nation's territory, including territorial waters, ensuring coordinated and effective responses in times of crisis.

2.3 Relationships between entities

The emergency and security communication needs of MCPTT users involve intricate information exchanges within each entity and between various entities, especially during joint actions and operations.

Inter-entity relationships are structured on two levels: firstly, within the context of their primary missions, categorised into four distinct areas (civil protection, internal security, civil emergency planning, and national defence); and secondly, through collaboration and cooperation with entities whose primary missions fall into the other areas.

Given this dual perspective and the fact that many entities have responsibilities spanning multiple categories, it becomes clear that interactions are pervasive among all entities. This is a fundamental requirement for MCPTT communications, ensuring interoperability and a minimum acceptable level of security trust for all entities.

Despite the high level of connectivity, it is crucial to ensure that command and control responsibilities, which support coordination functions, vary according to the specific framework of each action or operation. Given the unique nature of the four categories, it is possible for more than one entity to simultaneously hold command and control responsibilities for ongoing actions or operations.

As a final remark, PPDR systems must ensure reliable emergency and security communications for and between user entities during actions and operations within civil protection, internal security, civil emergency planning, and national defence.

2.4 Operational means

The operational means of the entities using MCPTT communication systems, for which the system must ensure emergency and security communication needs, are as follows:

- Agents and Operational Forces Groups/Teams;
- Vehicles:
    - o Terrestrial;
    - o Aerial (aircraft and drones);
    - o Maritime, in inland waters and territorial sea;
- Fixed Installations:
    - o Operational command and control centers of the user entities;
    - o Emergency and civil protection operations and communications rooms;
    - o Decision and monitoring centers of organisations responsible for forests, nature conservation, industry, energy, transport, communications, water resources, meteorology, geophysics, agriculture, sea, food, environment, and cyberspace.

2.5 Use cases

In terms of characterising operations and the subsequent requirements applicable to PPDR systems, the documentation published by the International Telecommunication Union (ITU), the Electronic Communications Committee of the European Conference of Postal and Telecommunications Administrations (CEPT/ECC), and the European Telecommunications Standards Institute (ETSI) is highlighted, especially:

- ITU-R Report M.2377-1 (11/2017)—Radiocommunication objectives and requirements for Public Protection and Disaster Relief [7];
- CEPT, ECC Report 199 (May 2013)—User requirements and spectrum needs for future European broadband PPDR systems (wide area networks) [8];

- ETSI TR 102628 V1.1.1 (2010-08)—Electromagnetic compatibility and Radio spectrum Matters (ERM); System reference document; Land Mobile Service; Additional spectrum requirements for future Public Safety and Security (PSS) wireless communication systems in the UHF frequency range [9];

- ETSI TR 102022-1 V1.1.1 (2012-08)—User Requirement Specification; Mission Critical Broadband Communication Requirements [10].

Given these considerations, PPDR systems should be prepared for operations in three types of scenarios [7–13], as follows:

- *Day-to-day or routine operations* (Scenario 1): These include the routine operations that PPDR entities conduct as part of their missions.

    - Typically, these operations occur within the national territory of the respective jurisdiction;

    - Generally, most spectrum and infrastructure requirements are determined based on this scenario, with the addition of extra capacity to meet unforeseen emergencies;

    - Examples: apartment/house fire, ambulance collection and transport of an injured person in a road accident, Christmas/Easter traffic control operation, street robbery, etc.

- *Major emergencies or large public events* (Scenario 2): Operations that occur in events not included in Scenario 1:

    - Events that public protection and sometimes disaster recovery authorities and organisations must respond to in a specific area of their jurisdiction while maintaining routine operations in other areas;

    - The scale and nature of the event may require additional PPDR resources from adjacent jurisdictions, foreign authorities, or international organisations;

    - In most cases, there are either existing plans or sufficient time to plan and coordinate the requirements;

    - Examples in Portugal: Papal visit in 2023, international football event (EURO 2004), national elections, international events (*e.g.* Web Summit), failure of public electronic communication networks, power or water supply disruptions, etc.

- *Disaster recovery* (Scenario 3): Special operations to minimise the effects of a disaster:

    - Disasters are situations caused by natural or human activity;

    - Generally, existing public protection communication systems and special theatre of operations communication equipment brought by disaster recovery authorities and organisations are used;

o Examples:

- Natural/other disasters: earthquake, storm, flood, large-scale forest fire, dam failure, volcanic eruption, tsunami, pandemic, etc.;

- Human-origin disasters: large-scale criminal incidents, armed conflicts, aircraft crashes, ship collisions, terrorist attacks on critical infrastructure, etc.

3. Mission critical communication network and deployment

3.1 Architecture and installed equipment

The architecture of the MCPTT communication systems network consists primarily of sets of switches and associated base stations, interconnected by redundant links with autonomous and independent operational capabilities. **Figure 2** illustrates an idealised set of base station configurations aimed at providing a resilient network. This example includes backhaul links utilising both microwave (B) and VSAT (A) links. These links serve as redundancy to the standard microwave (or fibre optical cable) (C) connections as default, ensuring enhanced reliability and robustness in the network infrastructure. In rural areas, microwave links should be adopted for reliability and cost effectiveness, as opposed to fibre optical cables deployed in underground conduits which tend to be very expensive, except for very short distances. On the other hand, overhead cables should be avoided, as they are exposed to extreme events or damage by natural or human causes. In the figure, the cell highlihted in red depicts the possibility of the existence of a radio exclusion zone for lack of radio coverage, where MCPTT communications are provided by a mobile base station with a satellite link.

The interconnected switches form "clusters." The MCPTT communication systems network, in terms of its geographical distribution, comprises several independent

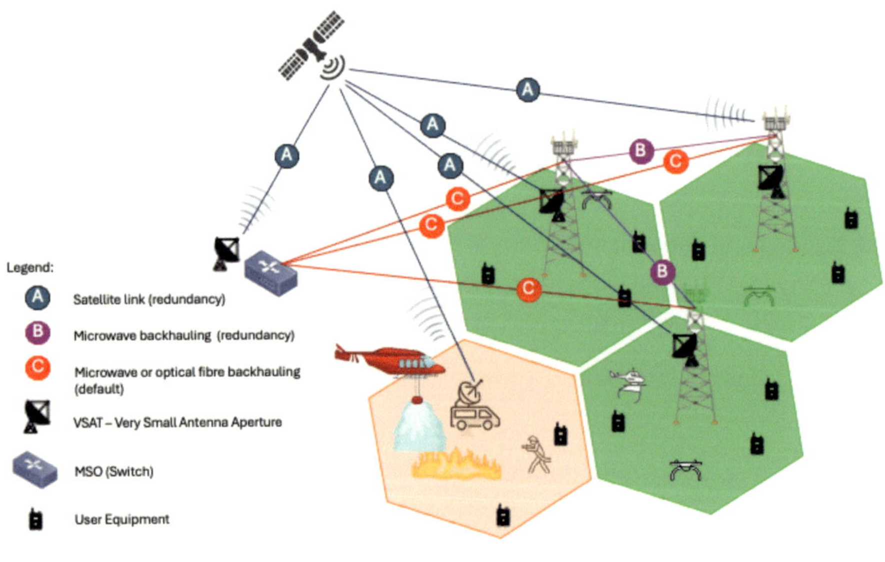

Figure 2.
Mission critical communication network with microwave link and satellite redundancies.

clusters. The connection between switches and base stations may use microwave links or fibre optic links in appropriate underground conduits. In addition to these, all base stations should have a redundant, possibly satellite (VSAT, very small aperture terminal) connection, ensuring resiliency.

The dispatch centers are the most powerful network elements accessed by users and are composed of the following:

- *Control room*: the Demarcation Device in the interconnection between the MCPTT communication systems network and the User Entity's domain. There should be one control room for each geographical location. Each control room is interconnected with its originating Mobile Switching Office (MSO), specifically its host zone;

- *Dispatch console*: consists of a workstation where (in the of TETRA) the Motorola dispatch application is installed, along with the respective audio interfaces—microphone and speakers. Each dispatch console is connected to its control room via an Ethernet connection, using the TCP/IP protocol.

The dispatch console ensures the controller's audio interface and, through the operator, enables a connection between the MCPTT end users and the remaining communication network, including the public switch telephone network (PSTN). The microphone captures and transmits control audio to the control room, which places it in the host zone, subsequently integrating it into the MCPTT communication systems network via the interconnected MSO. The operator listens to the return audio from the control room via the speakers. This way, audio interfaces act as format converters for the data circulating between the console and the control room.

3.2 Information system

The Information System for the TETRA communication network was developed based on the eTOM (Enhanced Telecom Operations Map) model with a specific extension – the Voice Recording System (VRS) [1].

A suite of applications and services supports the TETRA network operation, covering the following areas:

- *User support and service activation*: Ensuring seamless access and utilisation of services for users;

- *Support, management, and supervision*: Providing comprehensive management and oversight of the network's operations;

- *Decision support*: Facilitating informed decision-making through relevant data and analytics;

- *Voice recording system*: Capturing and storing voice communications for operational and compliance purposes.

It provides user support, service activation, management, supervision, and decision support. Key services include geo-referencing, which tracks terminals in real-time using GPS via the TETRA network, offering features like historical position visualisation and route tracking. Another service allows for real-time registration of road accident locations, detailing information flow between various data centers. Additionally, the system supports operational status messaging, which indicates the status of resources through pre-programmed messages on radio terminals. This messaging can be integrated with geo-referencing services for dynamic updates, enhancing the supervision and management of resources across different user entities.

3.3 Key quality indicators of land-mobile radio networks

Ensuring reliable and effective communication within the TETRA network requires several key quality indicators to be carefully monitored and evaluated. These indicators provide a comprehensive overview of the network's performance [14], addressing various aspects critical for maintaining emergency and public safety communications high standards:

- *Network coverage*: Generally meets contractual requirements, and a reference signal of −93 dBm should be considered in forest areas;

- *Accessibility*: Measures include not-established calls due to congestion or failures and interrupted calls from interference or network issues;

- *Call establishment time*: Evaluated by intracell communication time (PTT to Talk-Permit tone) and switch delay (call request to authorise action);

- *Availability*: Assessed in terms of operational and global availability, identifying downtime causes for critical and non-critical elements;

- *Audio quality*: Evaluated using PESQ based on the MOS, ensuring high speech quality for emergency communications;

- *Server availability*: High overall availability around 99.97%, accounting for minimal planned and unplanned maintenance down times. Application access remained stable throughout the year.

The quality indicators for the network encompass several key aspects critical for ensuring reliable and effective communication.

Network coverage should meet specified contractual requirements or service level agreements (SLA). The minimum level of the radio signal of −103 dBm is equal to the dynamic sensitivity of mobile terminals specified in the ETSI standard for TETRA systems. However, the signal level in a test procedure is generally measured under ideal conditions, with the terminal installed on a proper mount in a vehicle. In non-ideal use, there are additional attenuations, including attenuation from the user's body for the mobile terminal (between 5 to 9 dB), Rayleigh fading (4 dB), and log-normal fading (7 dB). On the other hand, it is important to note that coverage evaluation from measurements is usually carried out along communication routes. A measured value of −103 dBm on communication routes does not necessarily guarantee coverage in bush and forest areas. Therefore, it is an optimistic value, whose compliance does not guarantee effective radio coverage. The value of −93 dBm should be used in the simulation of coverage for appropriate radio planning.

Accessibility measures evaluate network efficiency, focusing on lost calls, which are queued due to congestion or failures. Interrupted calls are dropped due to interference or network issues.

Call establishment time, defined by the duration for intra-cell communication and switch delay, reflects the network's responsiveness and impacts user experience during critical communications. Availability is measured in operational and global terms: operational availability addresses failures affecting critical elements, while global availability considers failures impacting non-critical elements. Downtime causes are illustrated to identify common issues and areas for improvement, including values not typically included in standard metrics.

Audio quality, assessed using the Perceptual Evaluation of Speech Quality (PESQ) based on the Mean Opinion Scale (MOS), ensures high speech quality essential for clarity in emergency communications.

Server availability for support systems and application access is high, typically around 99.97%, accounting for planned and unplanned downtime. Comprehensive monitoring and improvement approach ensures the network meets the high standards required for critical communication services across various user entities.

3.4 Impact of wildfires on radio coverage

Wildfires can cause significant attenuation of radiowave signals, particularly in the very high frequency (VHF) and ultra high frequency (UHF) ranges, which are crucial for emergency communications during disaster recovery [1, 15–18]. One key factor contributing to this signal loss is the ionised plume created during vegetation combustion, which contains organic and inorganic particles, including alkali and alkaline earth metals (A-AEM) like potassium (K), calcium (Ca), and magnesium (Mg). These elements, with low ionisation potential, are released as free electrons and ions during combustion, especially in the flaming stage when temperatures exceed $500^{\circ}C$. The ionised plume behaves as plasma, which can absorb electromagnetic energy from radio waves, leading to signal attenuation.

The RESCuE-TOOL project [1] led by the authors was initiated to address the critical challenge of maintaining reliable radio communications during wildfire events, which are essential for effective disaster recovery and emergency response. Wildfires pose a significant threat to radiowave propagation, particularly in the very high frequency (VHF) and ultra high frequency (UHF) ranges. These frequencies are vital for emergency communication systems, such as push-to-talk radios (in MCPTT communications), used by firefighters and first responders during fire suppression efforts.

One of the primary factors contributing to signal attenuation in wildfire environments is the formation of ionised plumes. These plumes are generated during the combustion of vegetation and consist of organic and inorganic particles. Among the inorganic components, alkali and alkaline earth metals (A-AEM) like potassium (K), calcium (Ca), and magnesium (Mg) play a crucial role. These elements, due to their high concentrations in plant matter and low ionisation potential, release free electrons when exposed to the intense heat of a fire. This process leads to the formation of an ionised plasma, which interacts with electromagnetic waves and absorbs energy, resulting in radiowave signal loss.

The cold plasma model (CPM) was applied in this project to better understand and predict the behaviour of radiowaves as they propagate through the fire-induced plasma. Experimental and theoretical studies were conducted to simulate these effects in various fire scenarios. The project aimed to model the excess signal loss caused by wildfires, which is particularly critical in ensuring that emergency communication systems remain operational in areas affected by fire.

A key outcome of the project was the development of a real-time simulation framework designed to monitor radio coverage during wildfires [15–17]. This framework leverages digital elevation models to account for topography and vegetation, using established propagation models (such as ITM and ITU-R P833.5) to predict radiowave behaviour. The tool provides real-time coverage maps, identifying "radio exclusion zones" where signals are likely to be obstructed or lost due to the presence of fire-induced ionised plumes. This capability enables emergency response teams to adjust their communication strategies and equipment deployment, ensuring continuous and effective coordination in the field.

3.5 Large coverage areas - Portugal as a case study

Conducting an appropriate coverage area study is essential for ensuring that radio communication networks provide reliable and effective service across all targeted regions, including those most vulnerable or critical. This study's primary objective is to evaluate the network's coverage and signal strength to ensure it meets the required standards and offers adequate service [1].

As illustrated in **Figure 3**, the coverage area study consists of three main components (highlighted in blue): network coverage simulation, calibration using real measurements, and shadow area analysis. This structured approach ensures that the radio communication network is evaluated comprehensively, identifying areas where coverage can be improved to meet the required standards.

The first component of the study involves simulating the network's coverage across the study area. This simulation aims to predict the network's coverage by utilising reference propagation models that account for various environments, such as rural areas with high vegetation density, and open rural, suburban, and urban areas. The frequencies considered for this simulation are 380–383 MHz for the up-link and 390–393 MHz for the down-link. Essential data for this simulation include the locations of base stations, the type and height of antennas in service, and the emitted powers.

The calibration of the prediction model is the second crucial component. The objective here is to ensure the accuracy of the simulation models. This calibration is achieved using real-life measurements from specific drive tests and historical data from previous studies, such as measurements from legacy networks if available. The calibration scope in this example includes around two million measurement points along road routes across the country, representing approximately 5% of the national coverage area. This extensive calibration process ensures that the simulation models accurately reflect real-world conditions.

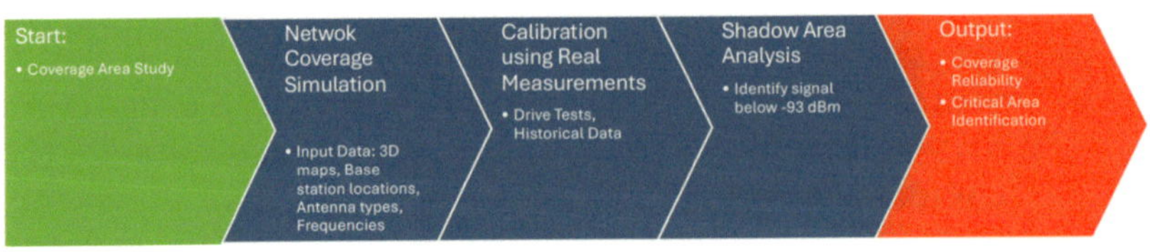

Figure 3.
Block diagram for conducting a coverage area study.

The third component of the study focuses on analysing shadow areas where the signal level falls below the acceptable threshold. The primary objective of this analysis is to identify and address areas where the signal level is below −93 dBm, which are considered shadow zones. The study aims for a coverage reliability of 99.47% in rural exterior environments, considering more than 500 base stations, as depicted in **Figures 4** and **5**, ensuring that even the most remote areas have adequate coverage. The latter includes the maritime coastline, extending 12 nautical miles (approximately 22.22 km). By including the maritime coastline, the percentage of covered area on the mainland Portugal shows a slight improvement. As a general recommendation, the service area of the emergency and security network should include the maritime coastline of the national territory, whenever applicable, to ensure adequate coverage and response capabilities.

In **Figure 6**, the simulation of the influence area of each base station is presented, considering the best signal from all base stations for mainland Portugal. The colours presented in the figure have no meaning other than to help identifying the influence of each base station.

Coverage is defined by the down-link, from the base station to the mobile station. This definition allows for more straightforward comparisons between the estimated values from the coverage calculations and the measured values. The Irregular Terrain (*Longley-Rice*) radio propagation model is used for this purpose, chosen for its effective calibration and accurate results, considering the varied orography of the territory. The spatial resolution for this analysis is set at 90 meters, providing a

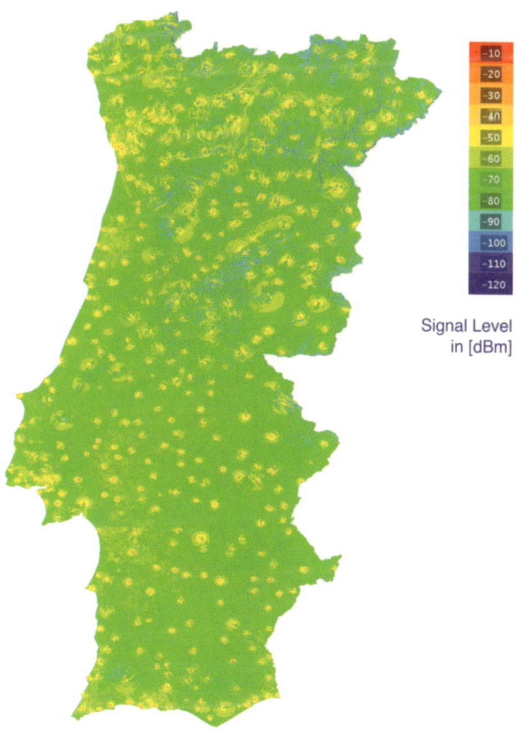

Figure 4.
Simulation of continental portugal radio coverage in 2019 at 393 MHz, considering the best server criterion at each location, excluding the 12 mile maritime zone [signal level, in dBm].

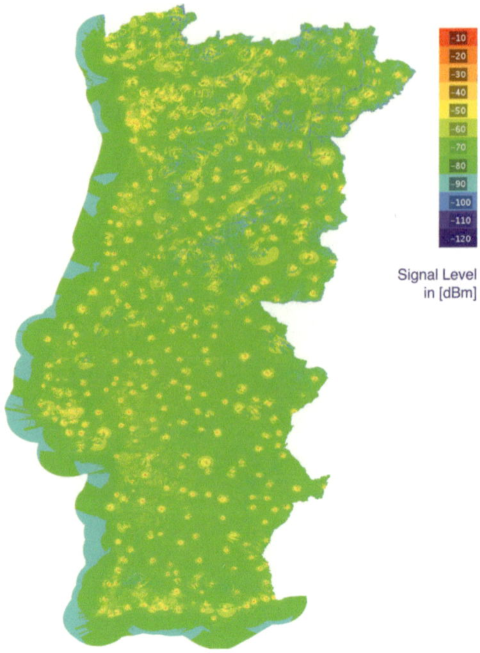

Figure 5.
Simulation of continental portugal radio coverage in 2019 at 393 MHz, considering the best server criterion at each location, including the 12 mile maritime zone [signal level, in dBm].

Figure 6.
Simulation of the influence area of each base station, considering the best signal from all base stations for mainland Portugal, excluding the 12 mile maritime zone. The colours have no meaning other than to help identifying the influence of each base station.

detailed terrain analysis. The median root mean square error between the measured and simulated signal intensity is 7.5 dB, ensuring a high level of accuracy in the study's findings.

To calculate coverage reliability, the study evaluates coverage by the best signal level from each base station at each location, known as the best server analysis. The territory is divided into 20 km by 20 km mosaics to facilitate detailed analysis and reliability calculations. Under these conditions, all areas should present a signal level above −93 dBm for 99.5% coverage reliability, under the specifications envisaged in the SLA. However, more realistic levels of −93 dBm and − 83 dBm are considered, especially applicable in rural areas more vulnerable to fires. The study finds that the percentage of covered areas reduces to 97.8% and 78.7%, respectively, when these more realistic levels are applied.

The results indicate high coverage reliability in the continental area, but they also identify critical regions where coverage enhancement is necessary. This identification of shadow zones and critical areas provides a basis for network improvements, focusing on areas with lower coverage to ensure comprehensive and effective radio communication. In conclusion, conducting a thorough coverage area study is vital for optimising network performance and ensuring reliable communications, particularly in critical and vulnerable areas. The methodology, including simulation, calibration, and shadow area analysis, provides a robust framework for evaluating and enhancing network coverage.

The study highlights significant issues with the MCPTT network's radio coverage in indoor environments, such as buildings and tunnels, which were not initially addressed. Despite attempts to extrapolate signal strength from the outside to the inside of buildings, the current standards do not ensure reliable coverage within structures, particularly those with high public occupancy or critical state services. Examples include shopping centers, stadiums, transport hubs, and hospitals. The lack of communication infrastructure in these locations poses a serious risk during emergencies, as response personnel may be unable to communicate effectively. Furthermore, existing legislation does not mandate that high-occupancy buildings implement MCPTT coverage, despite legal requirements for other safety measures, such as fire prevention. The study recommends extending MCPTT coverage to the interiors of public utility and state buildings, ensuring better preparedness in critical situations.

Based on the presented study, several recommendations are made to enhance the effectiveness and reliability of radio communication networks:

1. *Stringency of coverage requirements*: Although the current radio coverage satisfies contractual obligations, the existing reference level of −103 dBm is deemed insufficient for an emergency communication network. It is recommended to adopt a more rigorous threshold of −93 dBm or even −83 dBm. This change would significantly improve the likelihood of reliable radio coverage, particularly in critical situations where dependable communication is crucial;

2. *Strengthening coverage in key areas*: There is a need to enhance radio coverage in essential public infrastructure and areas with significant signal gaps (*i.e.,* poor radio coverage). This includes extending coverage to cross-border regions to ensure seamless communications. Strengthening coverage in these areas will enhance the overall reliability and effectiveness of the network, especially in serving public safety needs;

3. *Improved monitoring and reporting processes*: Effective, real-time processes for monitoring and reporting radio coverage should be implemented. Special attention should be given to remote areas far from main communication routes, as these are often the most challenging to cover. Tools that automatically collect and log coverage data from users' terminals in the field should be integrated into network management operations. Additionally, establishing an online portal for operations managers to access up-to-date information on radio coverage would be highly beneficial;

4. *Predefining mobile base station locations*: Optimal locations for mobile base stations should be studied and predefined for each region or municipality. This proactive measure ensures that competent authorities can quickly respond in emergencies. Predefining these locations will facilitate rapid deployment, enhancing network's responsiveness during critical incidents;

5. *Leveraging coverage overlap for capacity improvement*: Better utilisation of coverage overlap from multiple base stations, assuming no co-channel interference or controlled interference, can significantly increase the overall capacity of the system. By leveraging areas where cell coverage from more than one base station overlaps, the network can enhance both its capacity and efficiency. A more detailed analysis of this approach could further highlight the potential benefits and strategies for optimising these overlapping coverage zones;

6. *Extending coverage with vehicle-installed repeaters*: Consideration should be given to extending radio coverage by utilising vehicle-installed repeaters, which are effective in temporarily resolving coverage gaps in challenging scenarios. Conducting an inventory of such equipment already affiliated with the network could help identify untapped potential. Despite being underutilised, these repeaters could significantly improve coverage in hard-to-reach areas.

In conclusion, these recommendations aim to improve the radio network's coverage, reliability, and overall performance, ensuring it can effectively meet the demands of emergency communication and public safety.

3.6 Capacity analysis of the network: Insights from the Portuguese case

The Portuguese MCPTT network serves as an example of a communication system that was not designed with the flexibility to evolve in terms of coverage, service provision, or capacity to manage emergency situations. The initial traffic capacity was established based on population density, rather than accounting for potential risk areas or critical infrastructure.

This is in contrast to public mobile networks in Portugal and elsewhere, which continuously adapt to changing market needs and extraordinary events, such as large public gatherings or sporting events. Studies, including those conducted in Portugal, show that there is a significant overlap in coverage between neighbouring base stations, helping to reduce gaps and enhancing overall capacity.

To improve the capacity of the network, several steps can be considered, including increasing the number of base stations, adding more channels to each station, and refining the algorithm that mobile devices use to select base stations. In the current Portuguese system, base station selection is based purely on signal strength, which can result in devices connecting to overloaded stations or stations experiencing technical issues.

Based on the Portuguese experience, it is recommended to adjust this algorithm to enable mobile devices to switch to alternative base stations when the primary one becomes congested. This adjustment could greatly improve network efficiency during emergencies or periods of peak demand.

In Portugal, a geographic analysis of available radio channels per square kilometre highlights discrepancies between urban and rural areas. In rural regions, base stations typically cover larger areas but have fewer channels, making them more prone to congestion during crises. While the initial channel allocation in Portugal was based on local population densities, it is recommended that base stations in larger, rural areas receive additional channels to ensure they can support peak traffic in emergency situations. **Figure 7** shows the calculation of the density of available radio channels (cumulatively) per square-kilometre, based on point-to-point (pixel) best server analysis, for the entire mainland Portugal. For the calculation, the number of available channels at each base station was normalised by its coverage area at the reference level of −93 dBm, resulting in a median density of available radio channels of 6.78 $\left(\times 10^{-3}\right)$.

Additionally, when considering the influence area of each base station, it is observed that cells in rural areas have a larger coverage area, typically with only two assigned radio channels, making them more susceptible to traffic congestion during a crisis. It is also noted that base stations with two channels account for 86.11% (or 85.15% including the coastal area) of the radio coverage across the national territory.

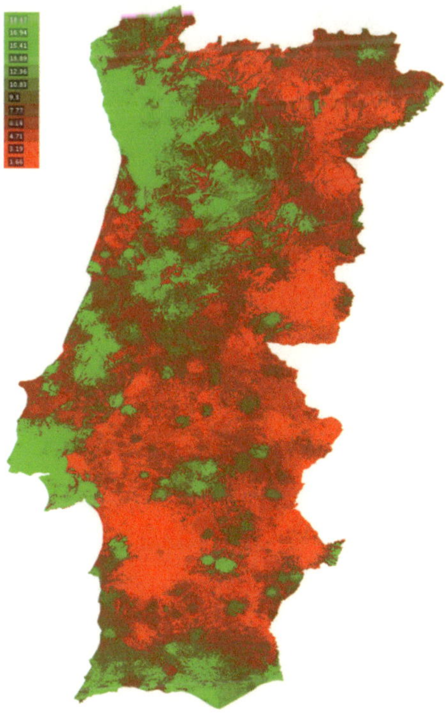

Figure 7.
Radio channel density per $km^2\left(x10^3\right)$ *in best server Mode in 2019 (Reference Level of −93 dBm), for Mainland Portugal.*

Despite the predefined criteria for channel allocation at each base station, which were based on the population density of the area being served, it is recommended that the number of channels be increased to a sufficient number (at least four) for base stations with larger area-to-channel ratios, to support peak traffic during a disaster or emergency.

3.7 Feasibility study of coverage in the 700 MHz band

Given the technological evolution of emergency and security networks, considering the recent international advancements in PS-LTE (Public-Safety Long-Term Evolution) standards, it is deemed appropriate and relevant to conduct a preliminary feasibility study of coverage in the 700 MHz frequency band, subject to prior authorisation from the competent authority (national regulatory bodies).

For the study, the same physical and radio assumptions of the current MCPTT network were considered, namely, the location and height of the base stations and effective isotropic radiated powers (EIRP), except for the frequency of the carriers, which were updated to 700 MHz. Thus, the study aimed only to assess the effect of propagation channel attenuation in the new frequency band.

The same reference levels were also used, although it is recommended that −83 dBm be considered more realistic due to the increased bandwidth (up to 10 MHz) and the signal-to-noise ratio (up to 13 dB) resulting from the use of higher-order discrete analog modulations (16-QAM) as recommended in long-term evolution (LTE).

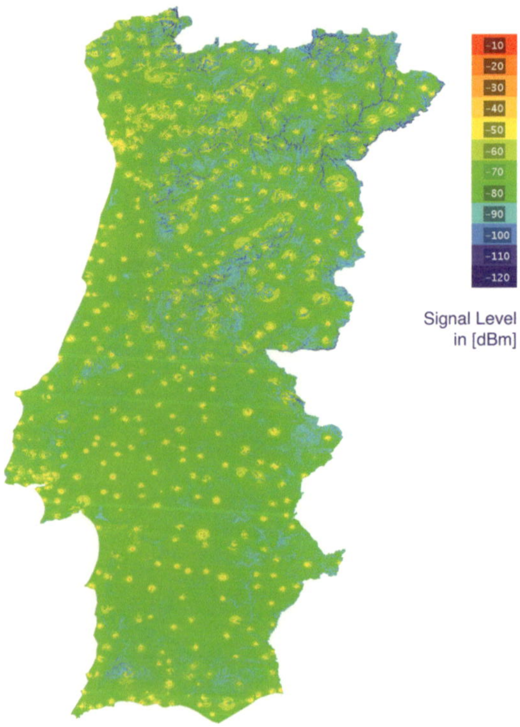

Figure 8.
Simulation of Mainland Portugal's coverage in the 700 MHz band, considering the best signal from all base stations.

The sensitivity of mobile terminals varies between −104 and − 94 dBm at the antenna input port, depending on the bandwidth and/or modulation order used.

Figure 8 presents the simulation of mainland Portugal's coverage in the 700 MHz band, for the best server scenario. A significant decrease in the covered area is observed, dropping to 97.77% for the reference level of −93 dBm. When considering the more realistic reference level for these frequencies (> −83 dBm), coverage reduces to 88.63%, which is remarkable.

The increase in shadow areas in the 700 MHz frequency band necessitates considering the coexistence of the new network with the current one.

3.8 Backhauling

3.8.1 Rationale for using microwave links in MCPTT networks

The use of microwave links in MCPTT networks offers several compelling advantages that make them an optimal choice for public safety and emergency response communications, as depicted in **Figure 9**. These links provide robust and reliable connectivity, essential to ensure uninterrupted communication during emergencies. They can be configured in redundant setups, such as bidirectional 1 + 1 systems with hot standby (HSB), which enhances their reliability. Microwave links also offer rapid

Figure 9.
Illustration of Reliable Wildfire Communication: Microwave links with 1 + 1 redundancy ensure uninterrupted, resilient connectivity between communication towers, enabling seamless coordination for firefighters in critical wildfire response scenarios.

deployment and flexibility, allowing for quick setup and reconfiguration, especially in challenging terrains and remote areas where laying fibre optic cables would be impractical. This flexibility makes them ideal for dynamic and evolving MCPTT network requirements. Additionally, microwave links are cost-effective, with lower installation and maintenance costs as compared to fibre optic cables. They support high data rates and provide the required bandwidth for data-intensive MCPTT applications, thanks to modern technology and the ability to use higher-order modulation schemes and various frequency bands. Furthermore, microwave links are scalable, allowing easy upgrades to support growing communication demands without major infrastructure changes. They also have a minimal environmental impact, as shown in **Figure 9**, making them suitable for protected or sensitive areas where large-scale construction is restricted. Overall, the reliability, flexibility, cost-effectiveness,

Figure 10.
Study of possible microwave links between base stations, up to a maximum distance of 20 km.

high performance, scalability, and minimal environmental impact of microwave links make them a highly effective solution to ensure robust and resilient MCPTT networks, providing emergency response and public safety officials with the reliable communication tools they need.

3.8.2 Interconnection of base stations and switches by microwave links

Given the Portuguese case study, a preliminary technical feasibility study for the interconnection of base stations to switches with redundancy by microwave links is presented. The study considered that the carrier frequencies to be used should be in the frequency bands of 3.5, 13, 15, or 18 GHz.

In the following example, the results of the links for 3.5 and 13 GHz are presented, in which, for the present study, only the existence of a radio line of sight with 100% (or 60%) clearance of the first Fresnel zone and up to a maximum link distance of 50 km was considered. For longer-distance links, the respective safety margins must be considered, according to the established international standards by the ITU-R, specifically ITU-R Recommendation F.1093-1, considering the effects of fading and rain.

Figure 10 presents the possible combinations for interconnecting base stations via microwave links, limited to 20 km. Assuming maximum link distances of 20, 35, and 50 km, it is observed from the analysis that each base station has, on average, at least two possible connections with neighbouring base stations. In some cases, depending on the maximum distance considered for the link, this number can reach up to 10 connections. This substantiates the feasibility of this solution, making it possible to find alternative paths to the switch, with traffic concentration points, to ensure link redundancy and thereby increase the resilience of the connections to the switch.

The costs involved with this solution depend on economies of scale and specific contracts with manufacturers. However, recent prices indicate that it is possible to make each of these links with bidirectional 1 + 1 systems with hot standby (HSB) for around 15,000 EUR (costs estimated in Portugal in 2019), including installation at pre-existing base stations. This initial investment would be supplemented by potential annual spectrum usage fees (which at current prices, considering an average distance of 22 km/link, would amount to 500 EUR/year). It is worth mentioning that this radio equipment may support the data rates of the links in case it is decided to increase the traffic capacity of the base stations. Costs indicated here are approximated costs in Portugal, in 2019.

4. Network resilience and optimisation

The vast majority of failures in the Portuguese MCPP network in 2017 can be attributed to one of two causes: (1) transmission network failures and (2) power supply failures at base stations [1]. These issues are due to each base station relying on a single backhaul circuit, which is particularly vulnerable in remote areas, where it is often installed on overhead lines, making it more exposed to damage. As a result, these failures are closely correlated with extreme events such as storms, floods, and wildfire because failures of the MCPP network became especially evident during the devastating wildfires in the summer of 2017, an analysis of the preceding five years reveals that these vulnerabilities had already been present. Wildfires are the most frequent type of emergency, particularly affecting transmission network circuits. However, storms tend to have a more severe impact on the network, as they disrupt a larger number of base stations due to both transmission circuit and power line failures. It is also worth noting that during the incidents of 2018 in Portugal, despite extended periods of failure, no downtime occurred thanks to the redundancy measures that had since been put in place.

Failures in the MCPP network can significantly affect the operational capacity of critical public services during emergency operations, such as emergency management agencies, fire departments, police forces, medical services, and the military, among others. These network failures, along with their frequent occurrence, are unacceptable. It is crucial to implement redundancy mechanisms for both the transmission network and power supply to ensure the reliability of the MCPP network in critical situations.

4.1 Installation of redundancy mechanisms in 2018 - Portugal as a case study

Until the end of 2017, the primary weakness of the Portuguese MCPTT network was the complete lack of redundancy in its transmission infrastructure and only partial redundancy in its power supply, which was limited to six hours of battery backup. Whenever transmission cables were damaged, for instance, by natural disasters such as wildfires, the connection between base stations and network switches was severed or disrupted. Similarly, when power lines were cut, base stations were limited to six hours of battery operation.

In 2018, for the first time since the network's deployment, redundancy solutions were introduced for both the transmission network and power supply, significantly improving the network's resilience to emergencies and disasters.

4.2 Transmission network - Backhauling

The creation of redundancy mechanisms for the MCPP network's transmission can be achieved in three ways:

1. A new alternative cable route;

2. Microwave links;

3. Satellite connections.

Transmission failures occur most frequently in remote areas where there are no conduits for fibre optic cables, and the existing routes are overhead on wooden poles, exposed to extreme events. Simply creating an alternative cable route would not fully resolve the issue, as the same extreme event could damage both circuits if they share the same vulnerabilities. Additionally, building conduits for these cables in such remote areas would be prohibitively expensive, and it is uncertain whether this would guarantee the complete resilience of the network, which would still be exposed to risks such as earthquakes, landslides or even fires at the base stations themselves.

Microwave links offer several advantages. This technology is mature, relatively low-cost, especially in terms of maintenance, and, while there is a slight risk of link misalignment, this is not a common occurrence. It is a particularly resilient solution to the extreme weather conditions that frequently affect the country.

In terms of data throughput, microwave links can easily support 100 Mbit/s connections, providing more than enough bandwidth for a TETRA voice network and even supporting a future upgrade to an emergency mobile network capable of handling data services and real-time video transmission, or both technologies simultaneously.

A solution based on microwave link systems, however, presents greater complexity and requires more time for planning. The study and budgeting of this solution necessitate analysis and evaluation for each base station, including assessment of the locations at each end of the microwave links, installation conditions, verification of line-of-sight, and/or the need for repeaters. This prevents a one-size-fits-all solution that can be applied to all network base stations. These systems always require elevated infrastructure, which in some cases may involve reinforcing existing communication towers or installing repeaters where there is no line of sight. The acquisition, planning, and implementation of a redundancy solution based on microwave links for a network like the Portuguese MCPTT network could take the government around two years to complete.

Lastly, a *satellite-based* redundancy solution for the transmission network is particularly effective because it uses a dedicated and contracted satellite segment. This ensures communication between base stations and the satellite station in case of transmission network failure, from where the traffic is routed to the corresponding MCPTT network switches. Although this solution has higher maintenance costs than a microwave link solution, due to the need to permanently lease satellite bandwidth, the installation is simpler. The setup is standardised for all base stations, and the satellite antenna can be installed on the communication tower (where feasible) or a separate pole. This streamlined installation process allows the deployment and roll-out to be completed in just a few months.

The bandwidth of this satellite solution will be limited, around 10 Mbit/s for the aggregate of base stations, due to the high cost of leasing satellite capacity. The satellite solution will not be able to serve as a redundancy option for a future upgrade of the TETRA network (due to the limited bandwidth and latency), and it is unlikely to be scaled to handle the simultaneous failure of a large number of TETRA base stations.

In April 2018, the Portuguese government chose the satellite redundancy solution. The time required for its roll-out was likely a key factor in this decision, as it was the only technical solution that could be implemented before the summer of 2018. To keep costs manageable, dynamic sharing of satellite resources was adopted, leasing satellite capacity sufficient to cover the simultaneous failure of around 60 base stations.

Although the satellite solution is not ideal from a technical or financial standpoint, it was understandable that, in the short term, it was the only option that could safeguard the public interest by ensuring the operational continuity of communications during the summer of 2018. Due to the time-consuming nature of planning and implementation, it is essential to initiate the procurement process for redundancy solutions as early as possible. Microwave links with dual paths should be adopted for base stations whenever feasible to enhance network resilience.

4.3 Power supply redundancy

Redundancy mechanisms for power supply in a network can be implemented in three ways: (1) additional batteries; (2) fixed generators with automatic start; and (3) mobile generators.

Most network sites are typically equipped with batteries that allow base stations to remain operational for a certain period (usually around six hours) during a power outage. Considering that such outages, often caused by major events like wildfires or storms, can take a long time to resolve, it is essential to assess how each of these solutions can enhance the network's overall resilience.

For example, in Portugal, data provided by SIRESP SA for the period from 2012 to 2016 showed approximately 520 cases of power supply failures. Of these, 366 (around 70%) lasted less than six hours, which was within the capacity of the existing battery backup systems. However, 63 cases (about 12%) lasted between six and 12 hours, and 91 incidents (around 17%) extended beyond 12 hours. On average, these longer outages lasted for 34 hours, highlighting the need for more robust power redundancy solutions in the network.

Assuming the analysis period and the number of occurrences are statistically relevant, the first conclusion is that the existing redundancy solution already addresses 70% of the power supply failures. On the other hand, it can be concluded that even if the battery backup capacity was increased to 12 or even 24 hours, the network would still be exposed to approximately 17% of power supply failures. This increase in backup capacity would only make the network resilient to an additional 12% of the failure cases.

Battery banks are an expensive solution with a short lifespan, as their performance degrades over time. As a result, they are associated with high costs for site visits, maintenance, and replacement. Additionally, battery banks are often targeted for theft due to their value. This creates ongoing operational uncertainty, as the state may only discover during a critical moment that the batteries have been stolen.

Fixed power generators would be a solution for the remaining 30% of power supply failures. Since these generators would be pre-installed and stocked with enough fuel to operate for several days, they could start immediately when a power outage occurs or when the existing battery backup runs out. However, fixed power generators with fuel have an even higher capital cost than expanding battery capacity and share the same issues regarding theft and operational uncertainty.

Both of the previous solutions also have an additional problem: because it is impossible to predict which locations will be affected by future power outages, they require installation at all sites in the MCPTT network, demanding a significant initial investment to make the network resilient to such failures.

The mobile power generator solution avoids the associated problems, as these generators can be pre-positioned, in a state of permanent readiness, in secure locations with good access. The presence of a six-hour battery backup provides a practical time window for mobile power generators to be transported to the affected base stations in case of a power outage, ensuring the continuity of their operation. The sites have been prepared for the immediate connection of a generator when needed, with the installation of a power switch and a socket for connecting the generator, allowing it to start functioning as soon as it is set up on site.

On the other hand, as simultaneous power outages across all network base stations are unlikely, the mobile generator solution, unlike the previously discussed options, can be scaled to cover a reasonable number of sites, making it the most cost-efficient option. Fixed power generator installations or battery life extensions cannot be scaled down, as it is impossible to predict where these solutions will be required. However, the mobile power generator solution will only ensure resilience for the number of stations it is designed to cover. In the event of an incident causing a power outage lasting more than six hours at a large number of base stations simultaneously, and if the number of affected stations exceeds the combined capacity of the mobile generators and the existing mobile stations, the remaining base stations will cease to function as long as the power failure persists.

In 2018, the Portuguese government chose the mobile power generator solution, establishing three permanent mobilisation centers in Vila Real, Coimbra, and Santarém, each equipped with six mobile generators. According to the national wildfire hazard map, the areas most exposed to wildfires are primarily in the Algarve and north of the Tagus River, while regions affected by storms are typically along the country's coastline.

4.4 Performance of redundancy solutions

4.4.1 Transmission redundancy

In 2018, two major events tested the Portuguese network's redundancy: a wildfire and a severe storm. During the wildfire, no failures in the transmission circuits occurred, making it less suitable for evaluation. However, the storm severely impacted the network, affecting 70 base stations. Despite these disruptions, the transmission redundancy solution performed effectively, as none of the affected stations experienced downtime, thanks to satellite-based backup systems, ensuring continued communication when other networks were down.

4.4.2 Power supply redundancy

The mobile generator solution was positioned in key locations to provide rapid support during power outages. The wildfire and storm both caused power failures, but mobile generators ensured there was no loss of service at the affected base stations. The wildfire incident demonstrated the readiness of the mobile generator teams, despite challenges with accessibility, as the generators were deployed and operational in time to prevent any service interruption.

For larger-scale events like the storm, the number of mobile generators proved adequate, but the analysis suggests that future preparedness could be improved by adding more generators, especially in regions harder to reach, such as the southern parts of the country. This would increase resilience in the event of widespread power failures across multiple sites.

4.5 Suggested follow-up actions toward network resilience

To ensure continuous improvement in network resilience, the following follow-up actions are recommended for emergency communication systems:

1. Implementation of redundancy solutions:

    a. *Geographic redundancy*: Establish geographic redundancy for critical network components such as switching offices. This ensures that communication remains operational even if a key node fails.

    b. *Transmission redundancy*: Develop backup routes for transmission networks, such as alternative cable paths or microwave links, to mitigate the risk of outages caused by infrastructure failures.

2. Comprehensive user training:

   a. Implement regular and updated training programs for network users. Ensure that key personnel are familiar with the latest features and functionalities of the communication system.

   b. Conduct real-world simulation exercises to reinforce operational skills, ensuring users are prepared to handle communications during emergencies.

3. Optimisation of network operations:

   a. *Cell re-selection*: Implement measures to optimise how communication devices switch between network nodes, ensuring seamless transitions during outages based on network availability and capacity.

   b. *Fleetmap optimisation*: Work closely with key agencies to optimise the configuration of communication channels, minimising the risk of congestion during high-traffic events.

4. Enhanced monitoring and reporting tools:

   a. Deploy a real-time network monitoring dashboard to track system performance, operational status, and coverage. This tool should provide critical insights during emergency scenarios and help manage resources efficiently.

   b. Ensure that the monitoring system is accessible to key agencies, allowing them to manage communication resources, identify issues, and adjust communication strategies as needed.

5. Strengthening power redundancy:

   a. Review and improve the power redundancy measures for base stations, ensuring that battery backups or mobile generators can sustain operations during extended outages.

   b. Establish additional mobile generator hubs in high-risk areas, to ensure rapid deployment when power outages affect multiple sites simultaneously.

6. Continuous system evaluation and updates:

   a. Regularly review and update network infrastructure based on technological advancements, changing operational needs, and emerging risks. This includes assessing system vulnerabilities and making necessary upgrades to maintain high levels of resilience.

   b. Develop long-term strategies for system upgrades, such as transitioning to next-generation communication technologies that offer improved resilience, capacity, and security.

7. Collaboration with key stakeholders:

   a. Foster collaboration between network operators, government agencies, and private sector partners to ensure transparency, data sharing, and joint efforts in enhancing network resilience. Item Engage with equipment and service providers to address any issues related to the network's physical infrastructure or operational limitations.

Implementing these follow-up actions will significantly enhance the resilience of emergency communication networks, ensuring reliable performance during crises and supporting critical operations. The focus should be on redundancy, training, optimisation, monitoring, and collaboration to maintain robust communication systems that are prepared for any challenges.

5. Management, operation, and security of the communication network

The management and operation of a communication network should be approached with both short- and medium-term objectives. In the medium term, over a one- to three-year period, management should focus on planning, system development, and strategic updates. This ensures the system remains aligned with user needs, operational requirements, and technological advancements. In the short term, management should include daily operational tasks that ensure the continuous functioning, security, and optimisation of the system, tailored to the specific demands of ongoing operations.

This section outlines a comprehensive approach to managing, operating, and securing a communication network to maintain resilience, security, and adaptability. It involves creating a framework for daily operational management, long-term strategic planning, security compliance, and ensuring redundancy and resilience across the network.

5.1 Management and operation

Effective management of the network requires both short- and medium-term planning. In the medium term, which spans one to three years, the focus should be on strategic development and updating the network to align with evolving user needs, operational demands, and technological trends. This includes creating a roadmap that anticipates future requirements and ensuring that the system remains relevant and capable of meeting operational needs, particularly for organisations involved in public safety, emergency response, and disaster recovery.

In the short term, daily management activities must ensure the continuous and secure operation of the network. These tasks should include monitoring system performance, maintaining network security, and optimising the network to meet the specific needs of ongoing operations. Short-term management also involves responding to immediate operational challenges and ensuring that the system is always ready to handle emergencies. A dedicated body should oversee both short- and medium-term objectives, ensuring that strategic plans are aligned with user requirements and operational tasks are efficiently performed.

5.2 Operational management

Operational management of the network should be financially and administratively autonomous to ensure flexibility in decision-making and resource allocation. This independence allows for swift responses to emerging challenges and ensures that the network can be maintained, updated, and secured without delay. The operational body should be tasked with maintaining system functionality under routine and emergency conditions, ensuring that the network can handle increased demand during major events, emergencies, or disasters.

This body must also ensure territorial coverage and interoperability, making sure that the network seamlessly integrates with other communication systems used by emergency and public safety entities. Regular updates to the network, based on technological advancements and operational needs, are essential to keep the system at peak performance. Additionally, operational management should include user training to ensure personnel are well-equipped to handle the system during crises. In short, the operational body must maintain the network's readiness and continuously optimise it to meet the demands of its users.

5.3 Security and resilience

A robust and secure communication network is critical, especially in emergency situations where lives may depend on reliable communications. The network should comply with international standards for emergency communication systems, ensuring it meets technical and operational requirements for public protection and disaster relief. International guidelines such as those from ETSI, ITU, and ISO should serve as benchmarks for the system's security, resilience, and preparedness.

In addition to international standards, the network should also comply with national regulations regarding the security and integrity of communication networks. This includes implementing strong security measures for critical assets, establishing risk management protocols, and ensuring that all operational vulnerabilities are addressed. Security incident response must be a key component of the system's resilience strategy, with protocols in place to manage and recover from any disruptions. Adhering to these guidelines ensures that the network can maintain operational continuity, even in the face of physical or cyber threats.

5.4 Network architecture

The architecture of the communication network must be designed with redundancy in mind. This ensures that if one part of the system fails, another can take over without disrupting operations. Redundancy should be built into all critical components, from radio access networks to core network infrastructure. By incorporating multiple transmission paths and backup systems, the network can continue to function in the event of equipment failure or damage caused by natural disasters.

In addition to redundancy, the system must protect its critical infrastructure against physical and cyber threats. Safeguards such as backup power, secure transmission lines, and physical barriers should be implemented to protect network assets. These measures are essential to maintaining the integrity and availability of the communication network during emergencies, ensuring that the system can support public safety operations without interruption.

5.5 Supply chain management

Resilience in supply chain management is crucial to ensuring that the network remains operational under all circumstances. The system must avoid reliance on a single supplier for critical components and services, as this could create vulnerabilities. Instead, products and services should be sourced from trusted suppliers, and redundancy should be built into the supply chain itself. Critical network components should undergo regular testing to ensure their reliability, and alternative suppliers should be available in case of disruptions.

Supply chain management also involves ensuring that critical assets are maintained and updated regularly. This includes conducting periodic assessments of system components to identify potential points of failure and addressing any weaknesses before they lead to operational issues. By ensuring a resilient supply chain, the network can continue to function smoothly even when external disruptions occur, further enhancing the system's overall reliability.

5.6 Interim conclusion

Ensuring the resilience and security of a communication network requires a comprehensive approach that includes strategic planning, operational autonomy, adherence to international and national standards, and robust supply chain management. By building redundancy into the network's architecture, securing critical assets, and maintaining operational flexibility, the network can remain reliable and efficient even during emergencies. With a focus on both short- and medium-term objectives, the management and operation of the network will ensure its continued alignment with user needs and technological advancements, making it a vital tool for public safety and emergency response operations.

6. Future trends in critical communications

This section explores the evolution of communication systems for critical applications, focusing on the future needs of public safety and emergency services. It reviews the development of TETRA, assesses current user requirements, and evaluates the role of newer technologies like LTE in addressing emerging demands for voice, data, and video communications.

6.1 Evolution of the TETRA system

The TETRA (Terrestrial Trunked Radio) system has been a cornerstone of critical communication networks since its introduction in 1989. Initially developed as a robust solution for public safety and emergency communications, TETRA has evolved to provide reliable voice services, GPS functionality, and push-to-talk capabilities across both urban and rural areas. Despite these advantages, TETRA's limited data transmission speeds have become a challenge as modern operational demands increasingly require high-quality video and data streams. This limitation has prompted a re-evaluation of TETRA's role and the exploration of alternative technologies to meet future requirements.

6.2 User requirements

The communication needs of key public safety forces, including police, emergency medical services, firefighters, and the general public, vary significantly depending on operational contexts. TETRA has proven effective in delivering voice communications with and without infrastructure support, and its integrated GPS capabilities allow for accurate geolocation of personnel. However, as operational demands evolve, there is an increasing need for higher bandwidth data services. Applications such as real-time video streaming and data analytics for situational awareness are becoming critical. While TETRA supports basic data transfer, its limited capacity means it cannot fully address these emerging needs, necessitating the consideration of more advanced solutions.

6.3 Available evolutionary alternatives

One of the most promising technologies identified as a successor or complement to TETRA is LTE. LTE's ability to deliver high-speed data transmission makes it ideal for supporting services like real-time video streaming, which is increasingly required for effective situational awareness in emergency operations. However, the cost of building a private LTE network for public safety purposes is often prohibitive. Countries like the U.S. and the U.K. have adopted hybrid models that leverage existing public LTE infrastructure while enhancing it with features like Quality of Service (QoS) and increased capacity to ensure priority access for emergency responders. These hybrid models have shown promise in balancing cost-effectiveness with the need for secure and reliable critical communications.

In this context, it is recommended that frequencies in the 700 MHz band be allocated for critical communication systems, while the state delays significant investment in LTE for public safety until the technology has matured sufficiently for reliable emergency applications.

6.4 Future communication systems for public safety—A critical review

While private LTE networks offer the highest level of control and security for public safety communications, they come with significant financial constraints. Public LTE networks, when augmented with prioritisation and security measures, provide a viable alternative. Countries like the U.S. (with FirstNet) and the U.K. (with ESN) have extended public LTE networks to include coverage in previously underserved areas, improving resilience and capacity during emergencies. Hybrid devices that combine TETRA for voice and LTE for high-speed data are already available, allowing for flexibility in cost-effectively addressing different communication needs.

The critical role of MCPTT voice communications in supporting emergency response operations has been widely studied and analysed in various research works. For instance, Brady and Roy [19] provide a comparative analysis of MCPTT services over legacy Land Mobile Radio (LMR) systems and LTE networks, highlighting the advantages and trade-offs in transitioning to more advanced communication systems. Höyhtyä et al. [20] explore how commercial mobile networks can be leveraged to support mission-critical communications, with an emphasis on licenced spectrum sharing and network slicing. Borsatti et al. [21] delve into network slicing architectures for mission-critical applications within the 5G framework, presenting real-world implementations and experimental validation. Jarwan et al. [22] and Santiago et al. [23]

further analyse the capacity and scalability of public safety networks and the evolution of critical communication services over LTE and 5G networks, providing insights into network dimensioning and resource optimisation. Finally, Solé and Augé [24] introduce the concept of mission-critical "bots" to enhance automated interactions in emergency operations, demonstrating the potential of innovative technologies in improving emergency response coordination.

The combined analysis of these studies highlights several key aspects and challenges related to the deployment and optimisation of MCPTT voice communications in emergency response settings. It underscores the transformative potential of MCPTT voice communications for emergency response, yet also brings to light several technical and practical challenges that need to be addressed for successful deployment. One of the primary strengths across these studies is the comprehensive exploration of MCPTT's role in enhancing the reliability and resilience of emergency communications. Most notably, the focus on utilising LTE and 5G technologies, as seen in Jarwan et al. [22] and Santiago et al. [23], reveals a clear trajectory toward next-generation communication systems capable of supporting the high bandwidth and low-latency requirements of mission-critical applications. The comparative analysis by Brady and Roy [19] between LMR and LTE networks offers critical insights into the benefits of transitioning to more advanced LTE-based systems while acknowledging the trade-offs in terms of performance and coverage.

Höyhtyä et al. [20] provide an innovative perspective on the use of commercial mobile networks for mission-critical communications, leveraging technologies such as network slicing and licenced shared access. However, while the potential benefits of such approaches are well-documented, the practical challenges of integrating these technologies into existing emergency communication frameworks are not fully addressed. Real-world deployment and the complexity of maintaining service continuity in extreme conditions, such as large-scale wildfires or natural disasters, are areas where further exploration is necessary. The experimental findings by Borsatti et al. [21] on the use of network slicing in a multi-domain scenario highlight the importance of service automation and slice isolation, yet the scalability and adaptability of these solutions in dynamic disaster scenarios remain to be validated.

The emphasis on network resilience and optimisation, as discussed in Solé and Augé [24] and the propagation studies related to wildfire effects on radio communications presented in the context chapter, is a crucial aspect of ensuring the dependability of MCPTT communications. The concept of mission-critical "bots" introduced by Solé and Augé [24] adds a new dimension to the discussion by suggesting the use of automated entities for information gathering and situational awareness. This approach could potentially reduce the cognitive load on human responders and enhance overall operational efficiency, but it also introduces new technical and operational challenges related to interoperability, system complexity, and the seamless integration of such technologies into existing communication networks.

One of the notable gaps across these papers is the limited focus on the real-world application and large-scale deployment of MCPTT solutions. While many of the studies, such as Borsatti et al. [21] and Santiago et al. [23], present detailed theoretical models and experimental results, they fall short in addressing the practical challenges of scaling these solutions to national or regional levels, especially in geographically diverse areas like Portugal. This was precisely the purpose of the preceding sections. The propagation characteristics and coverage requirements specific to large-scale wildfire fighting, as discussed in the context chapter, highlight the need for more in-depth studies that consider the unique challenges posed by such environments. For instance, the requirement for more than 500 base stations to cover Portugal at 400 MHz, compared to over 700 base stations at 700 MHz, demonstrates the complexity of balancing coverage, frequency selection, and infrastructure costs in real-world deployments.

In summary, while the referenced papers collectively provide valuable contributions to the understanding and development of MCPTT voice communications for emergency response, there are several areas where further research and real-world validation are needed. Future work should prioritise large-scale field testing, cross-comparative analysis of proposed solutions, and the development of hybrid network architectures that can adapt to varying conditions and requirements. Addressing these gaps will be essential for realising the full potential of MCPTT voice communications in enhancing the effectiveness and safety of emergency response operations, particularly in high-stakes scenarious such as large-scale wildfire fighting.

Moreover, future communication systems for public safety will likely combine a variety of technologies, such as TETRA, LTE-PPDR, GSM, and Wi-Fi, to ensure a high degree of flexibility and resilience. TETRA, with its well-established infrastructure and features, will likely continue to play a significant role in voice communications for some time, while LTE's higher data speeds will gradually take over data-centric applications. This gradual integration ensures that emergency responders have access to the most suitable communication tools for each scenario, whether it be voice, data, or video.

It is also recommended that further studies be conducted to explore alternatives for communication between aircraft and ground personnel, as current solutions do not fully address this need.

6.5 Emergency communication systems for Civilian populations

In terms of communications for the general public during emergencies, existing solutions largely rely on SMS through public mobile networks. However, in disaster scenarios where mobile networks may be compromised, alternative methods are required. A more resilient solution would be the use of FM radio broadcasts to disseminate emergency information. This method has been successfully used in other European countries and offers a reliable way to communicate with the public when mobile infrastructure is unavailable. It is recommended that frequencies in the FM band be allocated for emergency communications with the public, ensuring that critical information can still be delivered during disasters.

6.6 Final remarks

As communication networks for critical applications continue to evolve, a hybrid approach that combines the strengths of TETRA for voice communication and LTE-PPDR for data is recommended. This "system of systems" will allow emergency services to use the best available technology for their specific needs. Furthermore, integrating FM radio as a backup communication method for the general public ensures a more resilient overall network during times of disaster. Careful planning and phased implementation will ensure that the transition to more advanced communication systems is seamless and cost-effective, supporting the ongoing mission of public safety and emergency response.

7. Conclusions

Effective communication is the backbone of successful large-scale wildfire management and disaster response. This chapter has highlighted the critical role of mobile radio emergency communications, particularly push-to-talk terrestrial systems, in ensuring the safety and effectiveness of firefighting teams operating in remote and challenging terrains.

The analysis demonstrates the limitations of current public and private communication networks, particularly in low-density population areas where they are most needed during emergencies. As such, the need for specialised land-mobile radio systems that can operate in low-frequency bands (e.g., 400 MHz) becomes paramount. These systems not only provide broader coverage but also ensure reliable communication under extreme conditions.

Key challenges identified include the need for more resilient and redundant network infrastructures, especially in regions prone to wildfires. Mobile base stations and satellite backhauls should be pre-deployed based on risk assessments to ensure swift recovery of services when fixed infrastructure fails.

Future trends in emergency communication systems, including hybrid models that combine traditional TETRA radio networks with modern LTE technology, offer promising solutions. These hybrid systems can meet the increasing demands for real-time data, video streaming, and advanced situational awareness applications, all of which are essential for improving response efficiency and public safety.

In conclusion, enhancing the resilience, coverage, and capacity of mission-critical communication networks is essential for the effective management of wildfires and other large-scale disasters. Investments in new technologies, strategic placement of mobile assets, and robust backhaul infrastructure will help ensure continuous, reliable communication that can save lives and protect property during emergencies.

Acknowledgements

The authors would like to extend their sincere gratitude to the Instituto de Telecomunicações (IT), Instituto Superior Técnico (IST) and the Polytechnic Institute of Leiria (IPLeiria), for providing the research facilities. Also, they would like to express their sincere gratitude to the authors of the study "Report of the Working Group appointed by Dispatch No. 3458/2019 of March 2021 from the Secretary of State for Civil Protection and the Secretary of State for Communications" and other related reports produced in 2017-2018 by IT: Ana Aguiar, António Navarro, António Rodrigues, Carlos Fernandes, Fábio Ferreira, Francisco Gomes, Jorge Borreicho, José Sanguino, Manuel Pedrosa de Barros, Nuno Borges de Carvalho and Nuno Leonor. Their expertise, dedication, and collaborative efforts have significantly contributed to the successful completion of those reports.

Special thanks are due to the Ministry of Internal Administration (MAI) for their vital support and collaboration throughout the RESCuE-TOOL project. The partnership with MAI, including the General Secretariat (SGMAI) and SIRESP, SA, was crucial in granting access to valuable data and resources that significantly contributed to the success of this research.

The authors are particularly grateful to the Secretary of State for Internal Administration, Patrícia Alexandra Costa Gaspar, for facilitating this collaboration and authorising the use of SIRESP network data for RESCuE-TOOL project

(PCIF/SSI/0194/2017). Additionally, the authors would like to thank the Portuguese Foundation for Science and Technology (FCT) for their financial support, which made this research possible. They also wish to acknowledge the contributions of their research colleagues and students, whose dedication and efforts were essential in achieving the successful outcomes of this project.

This work was supported by FCT - Fundação para a Ciência e Tecnologia, I.P. under project reference UIDB/50008/2020, and DOI identifier https://doi.org/10.54499/UIDB/50008/2020.

Conflict of interest

The authors declare no conflict of interest.

Declaration

The disclosure of information in this chapter concerning the Portuguese Emergency Communication Network has been previously authorised by the Ministry of Internal Administration (MAI) within the framework of the RESCuE-TOOL Project (PCIF/SSI/0194/2017).

Author details

Carlos Salema[1,2†] and Rafael Caldeirinha[1,3*†]

1 Instituto de Telecomunicações, Portugal

2 Institution Superior Técnico, Portugal

3 Polytechnic of Leiria, Portugal

*Address all correspondence to: rafael.caldeirinha@ipleiria.pt

† These authors contributed equally.

References

[1] Project: Radio Coverage for Emergency Communication Systems to Operate under Critical Wildfire Environments TOOL [Internet]. 2024. Available from: https://www.it.pt/Projects/Index/4620 [Accessed: 16 September 2024]

[2] Leonor N, Faria S, Vala M, Caldeirinha RFS. A combined ITM and LITU-R model for enhanced radio coverage predictions of Mission-critical communications in mountainous vegetated terrains. IEEE Antennas and Wireless Propagation Letters. 2022;**21**(9):1777-1781. DOI: 10.1109/LAWP.2022.3179625

[3] ETSI TR 102 445 V1.1.1 (2006-10). Emergency Communications (EMTEL)—Overview of Emergency Communications Network Resilience and Preparedness

[4] ITU-D Study Group 2 6th Study Period 2014-2017 Question 5/2 Final Report—Utilization of telecommunications/ ICTs for disaster preparedness, mitigation and response

[5] 4G and 5G for Public Safety—Technology Options, TCCA, UK

[6] Public Safety Communications Evolution, January 2019, CISA, USA

[7] ITU-R Report M.2377-1 (11/2017). Radiocommunication objectives and requirements for Public Protection and Disaster Relief

[8] CEPT, ECC Report 199 (May 2013)—User requirements and spectrum needs for future European broadband PPDR systems (Wide Area Networks)

[9] ETSI TR 102 628 V1.1.1 (2010-08)—Electromagnetic compatibility and Radio spectrum Matters (ERM); System reference document; Land Mobile Service; Additional spectrum requirements for future Public Safety and Security (PSS) wireless communication systems in the UHF frequency range

[10] ETSI TR 102 022-1 V1.1.1 (2012-08)—User Requirement Specification; Mission Critical Broadband Communication Requirements.

[11] ETSI TS 102 181 V1.2.1 (2008-02) Emergency Communications (EMTEL)—Requirements for communication between authorities/organizations during emergencies

[12] Electronic Communications Resilience & Response Group (EC-RRG) Resilience Guidelines for Providers of Critical National Telecommunications Infrastructure (Updated August 2018)

[13] ITU-T Recommendation E.409 (05/2004)—Incident organization and security incident handling: Guidelines for telecommunication organizations

[14] Salema C. Microwave radio links: From theory to design. In: Wiley Series in Telecommunications and Signal Processing. Vol. 73. New Jersey: Wiley. ISBN 0471420263, 9780471420262, 474 p; 2003

[15] Vala M, Faria S, Coimbra P, Leonor N, Felício JM, Fernandes CA, et al. Effects of fire plumes from pine needles on small-scale fading in radiowave propagation. IEEE Transactions on Antennas and Propagation. 2023;**71**(4):3473-3484. DOI: 10.1109/TAP.2023.3238130

[16] Faria S, Vala M, Coimbra P, Leonor N, Felício JM, Fernandes CA, et al. Measurement trials and modeling

of radio propagation through small-scale fire using Pinus Pinaster fuel bed. IEEE Antennas and Wireless Propagation Letters. 2023;**22**(3):581-585. DOI: 10.1109/LAWP.2022.3218935

[17] Vala M, Faria S, Leonor N, Felício JM, Fernandes CA, Salema C, et al. Complex permittivity study of a Pinus Pinaster needle bed on fire. URSI Radio Science Bulletin. 2021;**2021**(378):14-21

[18] Leonor N, Fernandes T, Sanchez M, Caldeirinha RFS. A 3-D model for millimeter-wave propagation through vegetation media using ray-tracing. IEEE Transactions on Antennas and Propagation. 2019;**67**(6):4313-4318. DOI: 10.1109/TAP.2019.2905957

[19] Brady C, Roy S. Analysis of mission critical push-to-talk (MCPTT) services over public safety networks. IEEE Wireless Communications Letters. 2020;**9**(9):1462-1466. DOI: 10.1109/LWC.2020.2993996

[20] Höyhtyä M et al. Critical communications over mobile operators' networks: 5G use cases enabled by licensed spectrum sharing, network slicing and QoS control. IEEE Access. 2018;**6**:73572-73582. DOI: 10.1109/ACCESS.2018.2883787

[21] Borsatti D et al. Mission critical communications support with 5G and network slicing. IEEE Transactions on Network and Service Management. 2023;**20**(1):595-607. DOI: 10.1109/TNSM.2022.3208657

[22] Jarwan A, Sabbah A, Ibnkahla M, Issa O. LTE-based public safety networks: A survey. IEEE Communications Surveys Tutorials. 2019;**21**(2):1165-1187. DOI: 10.1109/COMST.2019.2895658

[23] Santiago A, De Cerio AD, Sanchoyerto A, Liberal F. Analysis of mission critical services radio access network capacity limitations over 5G. IEEE Access. 2024;**12**:6191-6203. DOI: 10.1109/ACCESS.2024.3350902

[24] Solé DV, Augé AC. A distributed man-machine dispatching architecture for emergency operations based on 3GPP mission critical services. IEEE Access. 2018;**6**:11614-11623. DOI: 10.1109/ACCESS.2017.2782760

Chapter 3

Fire Performance of Open-State Cavity Barriers for Ventilated Façades

Kate T.Q. Nguyen, Thomas W. Loh and Rajeendra Godakandage

Abstract

Ventilated façades are beneficial for improving buildings efficiency and comfort. However, the benefits are accompanied by additional risks of rapid fire spread through the ventilated cavity. To address the fire safety issue, ventilated façades incorporate open-state cavity barriers that allow airflow through under normal circumstances and react to compartmentalise the cavity to retard fire spread in the event of fire. Existing test methodologies cannot accurately assess the fire performance of open-state cavity barriers when applied to a full-scale façade system. They are often questioned on their suitability to assess effective closure during a fire. The existing test methods concerning the performance of open-state cavity barriers during fire have been reviewed to identify their limitations and discuss solutions to provide improved evaluation of fire performance. It is clear that existing test methods should not be used to assess the fire performance of open-state cavity barriers in ventilated facades. However, they can be used to compare the relative performance of different cavity barrier types and configurations. The parallel panel test arrangement was identified as a suitable test configuration to address some of the identified limitations. The performance evaluation criteria used by existing test methods are suitable for use in future test methods; however, the exclusion of failure within the first five minutes should be reviewed.

Keywords: fire rating, ventilated facade fire, cavity fire spread, facade fire test, cavity barrier test, intumescent activation

1. Introduction

Exterior façade systems are crucial to the performance, functionality and safety of buildings. There are various design options available including ventilated rainscreen facades [1, 2] which have gained popularity for their sustainability, high thermal and moisture performance and aesthetic appeal [3]. A representative ventilated façade configuration is presented in **Figure 1a**. The main feature of these façades is the cavity between the cladding and the insulation layers, which improves thermal and moisture performance [3]. In wet conditions, it reduces condensation and moisture penetration

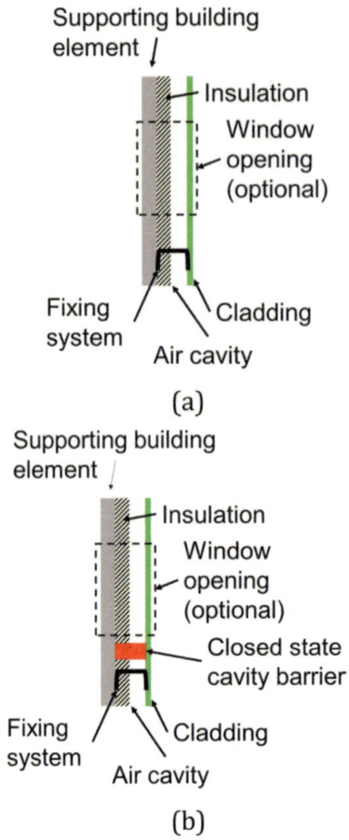

Figure 1.
(a) Schematic of a typical arrangement of a ventilated façade system with structural building elements, thermal insulation and cladding and (b) with a closed-state cavity barrier installed.

[4, 5], while during hot weather, solar radiation heats the cladding, causing upward air movement due to thermal buoyancy effects. The subsequent mixing of cool and heated air provides an insulation effect by transporting thermal energy out of the cavity [5]. This airflow mechanism is colloquially referred to as the "Chimney Effect" [1]. The facade ventilation openings can be located at the bottom, top, or at intermediate levels to enable air exchange with the outside environment [5], and the cavity width can vary from 20 mm up to 200 mm [5–8]. The cladding panels adjacent to the ventilated cavity are typically constructed using high-pressure laminates, sheet metal, ceramics, low-density polyethylene (LDPE), aluminium composite materials (ACM), timber, or cement and fibre-cement boards [6, 8–13]. The insulation layer creates the other boundary of the cavity and is typically constructed using expanded polystyrene (EPS), polyisocyanurate (PIR), polyurethane (PU), mineral wool, or expanded glass [6, 14]. The insulation layer also typically contains air and vapour barriers.

The fire performance of ventilated façades has been investigated [10, 11, 15–17]. For example, the fire performance of ventilated façades has been reported to be poorer compared to more traditional constructions [15]. In addition to the use of combustible cladding material, ventilated cavities were reported as an important fire safety consideration. During fire, the cavity can provide a low-resistance pathway for smoke and fire spread. Fire is most likely to penetrate a ventilated facade cavity through a broken window, for example, after a flashover event. However, when combustible cladding is involved in the fire, penetration into the cavity can also occur from outside of the building. Fire spreading through cavities can be disastrous due to the chimney effect combined with the use of combustible insulation and cladding materials (when present). It is known that the characteristics of cavity fires can be different compared to a typical compartment fire, and research has shown that the ventilated conditions can promote significant vertical flame spread [18].

In response to concerns over vertical fire spread, closed-state cavity fire barriers have been developed, **Figure 1b**. These barriers are typically fabricated from non-combustible mineral wool and are covered with aluminium foil or combustible polythene. Closed-state cavity barriers act to compartmentalise the cavity, thereby controlling the spread of fire to confined zones. A consequence of this strategy is the restriction of airflow within the cavity which compromises the heat and moisture transportation benefits. For ventilated façades to function properly, airflow within the cavity must be allowed during normal day-to-day conditions. However, during fire there is a need to close the ventilated façade to aid the control of smoke and fire spread.

Cavity barriers that allow the ventilated façade to remain open are broadly referred to as open-state cavity barriers. Open-state cavity barriers can be fabricated from the same non-combustible block material as conventional closed-state cavity barriers. To address the issue of smoke and fire spread, open-state cavity barriers can be fitted with the addition of a reactive mechanism to close the ventilated façade in the event of a fire. These mechanisms are typically fitted to the cavity barriers leading edge. Thermally activated intumescence substrates are exclusively employed as the closure mechanism to seal the cavity in the event of a fire. Examples of commercially available open-state cavity barriers with intumescent reactive closure mechanisms have been developed, see **Figure 2** [20–22].

A representative open-state cavity barrier with an intumescence coating before and during fire exposure is shown in **Figure 3**. Under normal conditions, a gap of 20 mm to 50 mm remains between the leading edge of the cavity barrier and the internal surface of the cladding, allowing airflow throughout the façade cavity. In the event of a fire, radiation from the fire or contact with flames heat the cavity barrier and activate the intumescent layer. The intumescent layer typically becomes thermally

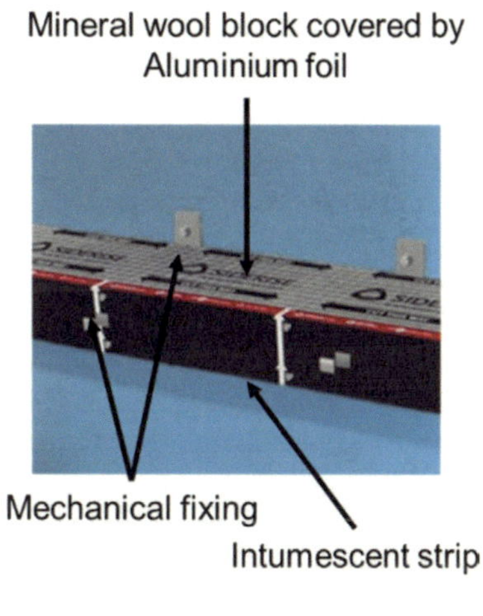

Figure 2.
Photograph showing the configuration of an open-state cavity barrier with an intumescent reactive closure mechanism [19].

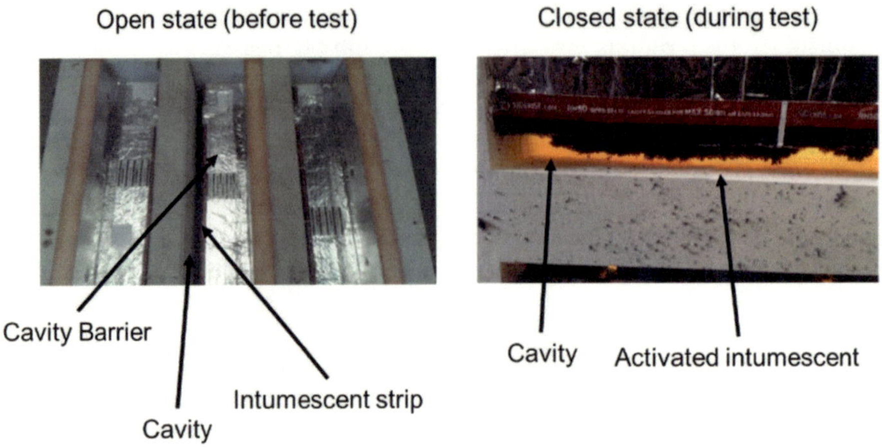

Figure 3.
Photograph of an open-state cavity barrier before thermal activation that allows airflow during day-to-day operation under normal temperature conditions (left) and after thermal activation upon fire exposure to close the cavity and restrict smoke and flame spread (right) [23].

active at around 200°C [19], although the exact initiation temperature varies between the different chemistries employed. Continued heating expands the intumescent coating and closes the gap, thereby restricting the spread of smoke and fire. It is important to note that open-state cavity barrier products exist with and without a reactive closure mechanism. For the purposes of this chapter, the term "open-state cavity barrier" refers solely to those with a reactive closure mechanism.

There is some uncertainty and, therefore, concern in the civil and fire-safety industry about the fire performance of open-state cavity barriers. Despite their commercialisation and implementation on existing buildings, there are no accepted standardised test methodologies to certify their effectiveness in controlling fire spread in ventilated façades. Test methodologies are being developed to evaluate their fire resistance and closure time; however, they are conducted on a small scale and do not necessarily include the façade system. Consequently, the results do not accurately represent the outcomes that can occur in a ventilated façade system during fire. This leads many to question the capability of existing test methods to evaluate performance. This chapter reviews the available test methods for open-state cavity barriers and aims to highlight areas of improvement that can lead to improved fire safety of ventilated façades under cavity fire spread conditions.

2. Existing test methodologies

The fire performance of ventilated façades can be evaluated using standard test methods such as ISO 13785 [24], BS 8414 [25], NFPA 285 [26] and SP FIRE 105 [27]. These test methods evaluate the overall performance of the façade system; however, their ability to evaluate fire spread through ventilated façade cavities is limited, often relying on a single temperature measurement inside the cavity to assess fire spread. They also do not directly evaluate the performance of open-state cavity barrier products. There are some important fire test methods that explicitly evaluate the fire performance of open-state cavity barriers used in ventilated façades, including TGD 19 [28], prEN 1364_6 [29] and ASTM E2912 [30], and these have been summarised and discussed below.

2.1 Technical guidance document 19 - Fire resistance test for "open-state" cavity barriers used in the external envelope or fabric of buildings

The Technical Guidance Document 19 (TGD 19) is a standard test method developed by the Association for Specialist Fire Protection (ASFP), United Kingdom [28]. The procedure assesses the fire resistance of open-state cavity barriers by considering a fire exposure beneath the cavity barrier and evaluating its effectiveness to seal and restrict the spread of fire. A cross-sectional view of the TGD 19 test assembly is shown schematically in **Figure 4a**. Two lightweight autoclaved aerated concrete leaves are constructed to create a cavity placed on top of a furnace. The open-state cavity

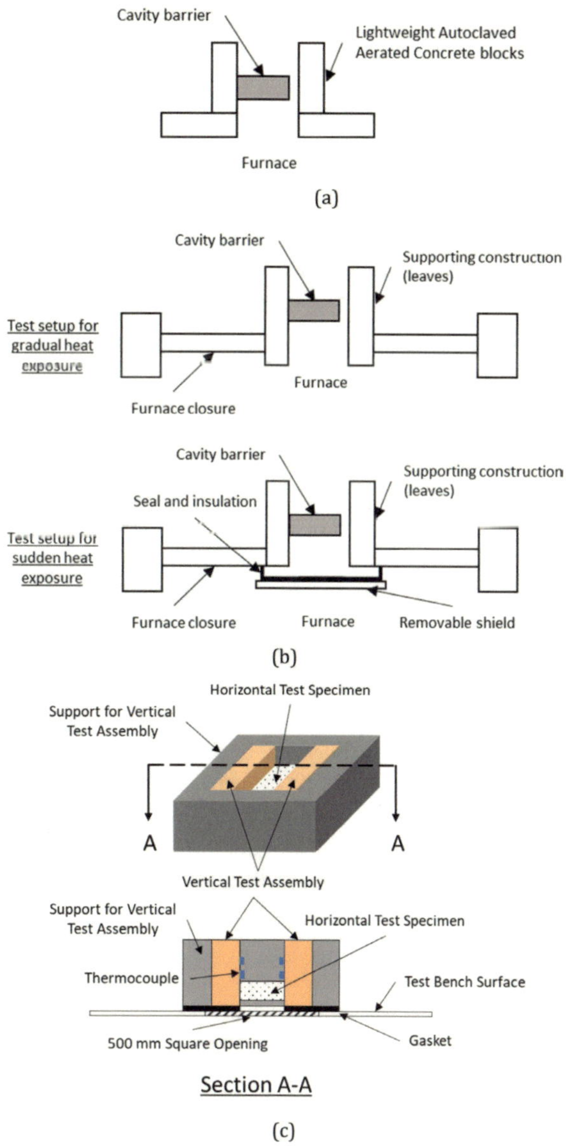

Figure 4.
Schematic representations of open-state cavity barrier testing arrangements for (a) TGD 19 [28], (b) prEN 1364_6 [29] and (c) ASTM E2912 [30].

barrier is arranged between the leaves along with substrate material representative of those used in the cladding construction. The fire condition is applied as a standard cellulose fire curve. Due to the time required to establish the specified temperature and pressure conditions, any insulation or integrity failure within the first five minutes of heat exposure are excluded from the assessment.

2.2 EN 1364_6 - Fire resistance tests for non-loadbearing elements

Currently, there is no active European Standard to assess the fire performance of open-state cavity barriers; however, a standard test method designated prEN 1364_6 is being developed [29]. At the time of writing, prEN 1364_6 is addressing public comments. Its evaluation is based on the principles used in the TR 31 technical report by European Organisation for Technical Assessment (EOTA TR31) [31]. However, EOTA TR31 only considers closed-state cavity barriers, whereas prEN 1364_6 considers both open- and closed-state cavity barriers. For open-state types prEN 1364_6 proposes to define a test procedure separate from their closed-state counterparts. The proposed test assembly is similar to TGD 19 and is presented in **Figure 4b**. The open-state cavity barrier is located between two leaves and is exposed to standard cellulose fire conditions using a gas furnace. However, unlike TGD 19, two approaches to fire exposure are allowed. The first approach exposes the cavity barrier to the fire condition as soon as the furnace is ignited. The second approach isolates the cavity from the fire for the first five minutes using a non-combustible removable heatshield. The heatshield is positioned between the fire chamber and the cavity and is removed after the first five minutes of the test to allow fire penetration into the cavity, simulating the sudden failure of materials surrounding the ventilated façade cavity.

2.3 ASTM E2912 - Fire test of non-mechanical fire dampers used in vented construction

The American National Standard ASTM E2912 [30] evaluates the effectiveness of non-mechanical fire dampers used in vented construction during the open state to limit smoke and fire spread. The test assembly is shown in **Figure 4c**. Contrary to the other test standards, ASTM E2912 employs a gas burner to produce the fire condition meaning direct flame impingement is included in the evaluation of the cavity barriers performance. This testing procedure is not specifically for the open-state cavity barriers with an intumescent reactive closure mechanism but generally for all non-mechanical fire dampers in vented construction. This variation in the scope of application is the main reason for the significant variation in the test setup compared to the other two approaches.

3. Alternative test methodologies

A possible alternative test arrangement for the evaluation of open-state cavity barriers is the parallel panel apparatus. This test arrangement has seen use in ANSI/FM Approvals 4880-2017 [32] and ANSI/FM Approvals 4411-2020 [33]. These standards evaluate the fire performance of insulated building panel assemblies and cavity wall systems, respectively. They consider the fire scenarios of fire ignition within the cavity from hot work, electrical shorts, and, importantly, fire extending through a window due to a flashover fire inside a room. The test consists of two parallel panels

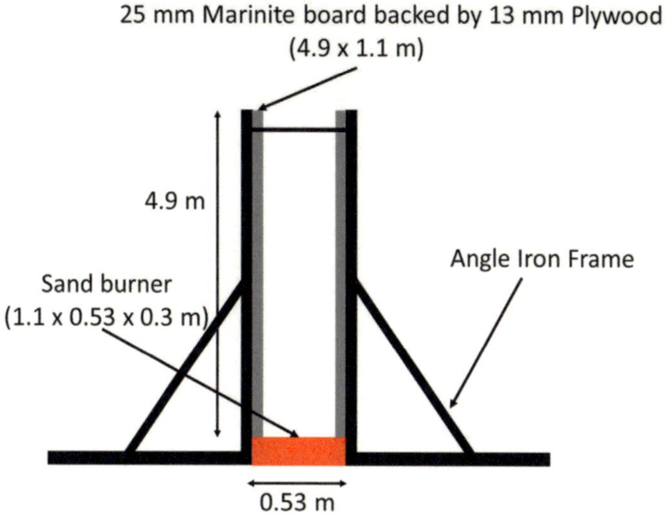

Figure 5.
Schematic example of a parallel panel test setup used in ANSI/FM approvals 4880-2017 [32] to evaluate the combustibility of materials used inside cavity spaces.

placed vertically with a cavity gap, **Figure 5**. The fire condition is generated using a propane sand burner positioned at the bottom of the apparatus. The cavity widths used in FM 4880 and FM 4411 are, respectively, 530 mm, and 50 mm or 100 mm, which account for the typical range of cavity widths required to accurately represent ventilated façades (20 mm to 200 mm). The height of FM 4880 (4.87 m, 16 ft) and FM 4411 (2.43 m, 8-ft) are also more representative of real ventilated façade cavities compared to those used in existing test methods. However, the length of the cavity is shorter at 1100 in FM 4880 and 1200 mm in FM 4411.

4. Discussion on appropriate test methodologies

Comparison of the test methods summarised in Sections 2 and 3 highlight several differences in the existing approaches, including application, fire exposure conditions, test setup, test conditions, instrumentation and performance criteria. Discussion on these topics and suggestions to further develop a standardised test method to evaluate the fire performance of open-state cavity barriers are provided in this section.

4.1 Field of application

TGD 19 and prEN 1364_6 allow the direct application of the test results for certain variations of the test setup. This is useful when alterations or variations must be made to the cavity barrier design after conducting a full-scale façade test. Such alterations include:

- Increases to the length of the cavity barrier when the specified minimum length is tested.
- Decreases to the distance between two fixing points.

- Interpolation of cavity width (distance from the insulation to the cladding) within the tested range.
- Decreases to the distance between the leading edge of the cavity barrier to the cladding internal surface, maintaining the minimum distance between the combustible insulation and the leading edge of the cavity barrier.

However, a significant disadvantage is that neither TGD 19, ASTM E2912, nor the proposed prEN 1364_6 can provide a reliable indicator of the fire performance of the cavity barrier when installed at full-scale. This is a concern explicitly noted in all approaches. For example, it is possible for a cavity barrier to exhibit a high level of flame spread prevention according to these methods but fail to achieve a similar level of performance when involved in a full-scale ventilated façade system fire. While the existing test methods can provide an indicative level of performance, their application must be limited to providing a relative performance assessment of different cavity barrier designs and products. Consequently, the existing test methods become redundant in validating the fire performance of a ventilated façade design through full-scale tests. At this stage, the requirement of a rigorous test standard that can conservatively predict the performance of an open-state cavity barrier is essential.

4.2 Test arrangement

It is important that the test arrangement appropriately represents the construction of the façade to accurately capture the fire dynamics within the cavity. The geometric parameters (length, width, height) of the cavity, combustible nature of the materials present in the façade and the overall construction (e.g., mounting systems) all influence the fire dynamics within the cavity and therefore the evaluation of a cavity barriers fire performance.

4.2.1 Geometric dimensions of the cavity construction

The width of the cavity in a ventilated façade can range from 20 mm up to 200 mm. For the test arrangements presented in **Figures 4** and **5**, the width of the cavity can be matched to the desired dimensions. The likelihood of fire penetration into a cavity is dependent on the cavity width [18] as well as the trajectory of the flame plume outside of the cavity [37]. Generally speaking, the larger the cavity width, the easier it is for fire to enter the cavity. The length of a ventilated façade can be comparable to the length of the building. Unlike the cavity width, it is not practical to match the test arrangement to the length of a building. The TGD 19, pr EN 1364-6 and ASTM E2912 test methods, respectively, have a maximum length of 3 m, 3 m and 0.5 m. With the exception of ASTM E2912, the lengths agree well with those used in other intermediate- and large-scale façade fire test methods that are required to replicate complex fire dynamics [25–27]. The height, like length, cannot be practically matched to the dimension of the building and a representative portion of the façade must be selected for evaluation. Cavity heights of 0.6 m and 0.5 m are used in the TGD 19 and prEN 1364_6 test methodologies, respectively. While in ASTM E2912 the height is 0.315 m (for a cavity barrier depth of 120 mm).

The geometric construction of the test arrangement has implications for the combustion and fire dynamics that occur inside of a ventilated façade fire. Whether or not the dimensions of existing test methods are sufficient to accurately evaluate the performance of open-state cavity barriers is still under question. It is important that the cavity width can match those used in practice as after fire enters the cavity the temperature and fire dynamics within the cavity are strongly influenced by the cavity width. For example, the temperature increases with decreasing cavity width for cavities constructed as non-combustible. For combustible cavities, the relationship is the opposite. The level of air entrainment is also affected by cavity with, reducing with decreasing cavity width [7, 35]. Since combustion is dependent on air entrainment, larger width cavities can have higher temperatures compared to narrower ones. Concerning the height, it is very likely that the existing heights are too short to establish the cavity fire dynamics necessary to be representative of an actual ventilated façade cavity fire. Realistically, flames can propagate well above the 0.315 m to 0.6 m heights used in the current test methods, and the cavity construction in this region can influence the fire dynamics below. The cavity height also influences heat loss and radiative feedback, and therefore the overall temperature within the cavity. Consequently, the geometric construction of existing test methods is likely not appropriate to evaluate cavity barrier performance.

4.2.2 Auxiliary cavity constructions

The three test methods stipulate the requirement to consider the substrates that match the actual application used during the test. The test usually incorporates insulation materials as well as timber, concrete and masonry substrates as necessary. However, clear indications of the requirement to use the cladding material are not provided. Due to the strong influence of combustible materials on fire spread properties and heat generation, it is crucial that their volatile nature is reflected in the assessment of open-state cavity barriers, both in the open and closed states. For example, flame heights of up to 1.4 m were measured for a cavity fire with a cavity width of 100 mm, heat release rates of up to 16 kW, and a non-combustible construction. When combustible materials are included within the cavity, the flame heights can be significantly higher, exceeding the confines of 2.4 m tall, ventilated façades test arrangements [6, 14, 36]. For example, flame heights in excess of 2.4 m were measured when combustible extruded polystyrene was included in a ventilated cavity fire test using parallel panels, **Figure 6**.

When considering the evaluation of façade systems, their in-plane construction can often be neglected or simplified. Features such as aluminium rail used to fix many cladding systems to the structural wall are an important inclusion for cavity fires. The rail systems locally consume a significant portion of the cavity width. During a cavity fire, the rails influence the oxygen conditions within the cavity by restricting air entrainment into the hot fire plume within the cavity. However, as the fire progresses, the aluminium can melt, generating openings that allow higher levels of air entrainment and aid combustion. It is important that the façade system as a whole be considered when evaluating the performance of open-state cavity barriers.

4.3 Specimen construction

Cavity barriers are fabricated in discrete lengths, up to 1.2 m. When installed on a building, multiple cavity barriers are spliced together to provide continuous protection along the desired length. This results in multiple joints along the cavity barrier which may act as weak points to initiate failure during fire. The robustness of the cavity barrier can also depend on the number of support locations used to fix the

Figure 6.
Example of the fire severity that can occur when combustible materials are included in the cavity of a ventilated façade depicted in a parallel panel test [14].

barrier to the wall. Sufficient specimen length should be considered to evaluate the splices and attachment methods. Representative cavity barrier joints and supports are presented in **Figure** 7.

TGD 19 and prEN 1364_6 have considered the application of specimen configuration by allowing sufficient specimen length (up to 3 m) to assess splices and the configuration of supporting locations. Therefore, these two approaches facilitate practical assessment of the joints between multiple specimens, considering that they resemble the actual façade application. ASTM E2912, however, only allows a maximum specimen length of 0.5 m, which cannot be used to evaluate the performance of joints and splices.

4.4 Test conditions

4.4.1 Fire scenario

Standard test methods for fire resistance of building elements must incorporate fire exposure conditions representative of the actual fire event. However, the fire exposure conditions used in open-state cavity barrier tests are often questioned regarding their resemblance to actual fire conditions within ventilated façades. A reference fire scenario for open-state cavity barriers has not been defined. Both TGD 19 and prEN 1364_6 rely on the fire exposure corresponding to a compartment fire scenario. However, it is known that fire conditions within the cavity of a ventilated façade differ from those of a compartment fire [18]. The combined influence of the chimney effect and radiative heat feedback from the cavity boundaries can induce

(a)

(b)

Figure 7.
(a) Photograph showing the joints of a typical cavity barrier system and (b) supports used to fix cavity barriers to walls [20].

a more severe heating condition than a standard cellulose fire curve exposure. For example, the early stages of fire growth in a combustible ventilated façade cavity fire have been reported to be more accurately represented by the standard hydrocarbon fire curve rather than the cellulose fire curve, **Figure 8** [37, 38]. Comparison of the standard fire curves and the measured values reveals the temperature to exceed the hydrocarbon curve by over 150°C in some instances.

The heating condition used in ASTM E2912 takes a different approach to representing fire exposure. In this standard, a 300 kW burner is used to apply fire exposure. The test attempts to replicate a fire scenario where a flashover fire inside a compartment results in flame ejection through the window housing and into the ventilated façade cavity. Using a burner allows the instantaneous and direct flame interaction with the cavity barrier from the beginning of the test. The approach appears closer to reality compared to TGD 19 and prEN 1364_6. For example, prEN 1364_6 allows the penetration of heat directly into the façade cavity *via* a removable heatshield to simulate the failure of the window housing. However, it does not account for potential direct flame interaction with the cavity barrier closure mechanism, nor the potential fire spread over combustible insulation and/or cladding material. Nevertheless, whether the heat release rate prescribed in the ASTM standard accurately represents all cavity fire penetrations has not been resolved.

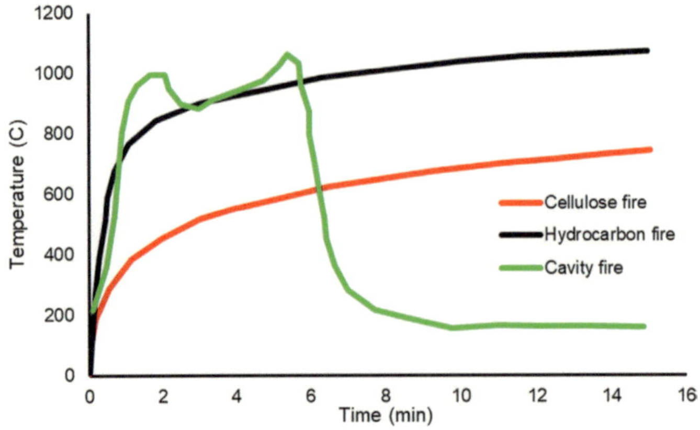

Figure 8.
Comparison of cavity fire temperature with standard fire curves [37].

Research has reported the use of the parallel panel test to evaluate the fire performance of ventilated façade configurations and other cavity types including external wall systems. For example, a FM 4880 parallel panel setup using a propane heat release rate of 360 kW was reported to induce a peak heat flux of 100 kW/m^2 on to the cavity panels [39]. When compared with peak heat flux values from façade fire tests, this value exceeds the equivalent measurements in the NFPA 285, CAN/ULC S134, ISO 137-85, DIN 4102-20, SP-105 and BS 8414 tests standards (**Figure 9**). The higher heat exposure in the parallel panel test is due to the high radiative heat flux resulting from the cavity geometry. Clearly, it is not accurate to stipulate the same heat exposure conditions as other façade fire tests. The current research appears to be an encouraging indication of the suitability of the parallel panel test to replicate the fire conditions that occur within cavity fires [40].

Research using the parallel panel test arrangement has demonstrated the importance of representing the geometric and combustible nature of the ventilated façade

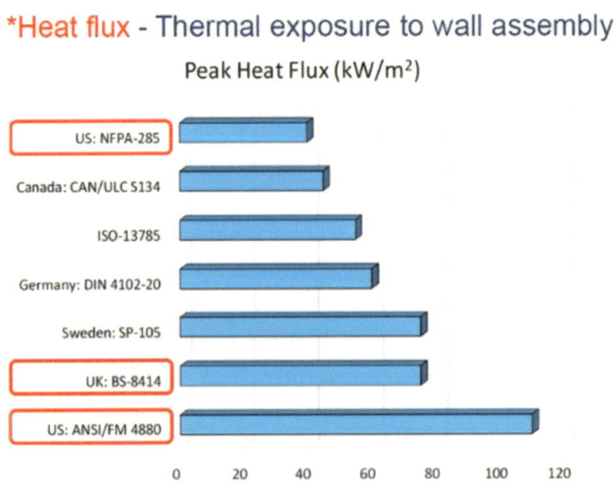

Figure 9.
Comparison of the maximum heat flux exposure measured during a parallel panel test and other standard façade fire tests [40].

to ensure the cavity fire behaviour is accurately captured during the evaluation of the cavity barrier's fire performance. The research to date provides a strong indication that the parallel panel setup may well represent the thermal conditions that occur with the cavity during ventilated façade fires, highlighting their potential suitability for evaluating the fire performance of open-state cavity barriers.

There is a need for a reference cavity fire heat exposure for open-state cavity barrier testing and this must consider the physical and combustible parameters of the cavity. Not all of these parameters are adequately considered by the existing test methodologies. One possible solution may be for the burner to mimic the temperature and pressure conditions measured at the bottom of the cavity during a compartment fire following a standard cellulose curve. This should generate temperature and pressure conditions similar to those beneath the cavity during flashover in a compartment fire. The subsequent flame penetration and growth within the cavity will then depend on the cavity's geometric construction and combustibility of its boundaries which will also influence the performance evaluation of the cavity barrier.

4.4.2 Pressure

Pressure is an important parameter when evaluating fire behaviour and this is true for ventilated façade cavity fires. During fire, pressure influences fire dynamics, flame spread behaviour, smoke movement and the integrity of building components (including cavity barriers). For example, the pressure differential across a fire plume is the driving force determining its trajectory in a façade cavity fire, and this depends largely on the relative distances between the plume boundary and the interior or exterior cavity surfaces, **Figure 10** [34]. This indicates that flame interacting with the cavity barrier could occur in different ways and durations depending on the pressure differential across the cavity width. As another example, a narrow cavity width combined with the chimney effect may create pressure conditions that differ from compartment and open fire scenarios [18]. However, to what degree does it requires further investigation? The pressure conditions surrounding or within cavities during a ventilated façade fire require further investigation to develop a prescribed requirement.

TGD 19 and prEN 1364_6 specify a pressure of 20 Pa, 100 mm beneath the cavity barrier. However, this pressure condition is not derived from cavity fire research; rather, it is based on a typical compartment fire scenario. A pressure condition is not stipulated in ASTM E2912 or the parallel panel tests, although the burner will induce pressure to some degree. The advantage of the parallel panel test is the physically larger configuration which enforces fire entering the cavity and fire-driven flow. An accurate pressure condition and the critical factors influencing it remain to be determined.

4.4.3 Flow velocity

The flow velocity within the cavity is another crucial factor when evaluating the fire performance of open-state cavity barriers. High flow velocities promote the upward spread of flames which interact with the cavity barrier and its closure mechanism. The high-velocity upward flow occurs due to the cavity being heated far beyond ambient temperatures, resulting in a drastic decrease in density, which allows the buoyant flows of the chimney effect to initiate and entrain fresh air into the cavity. Inadequate flow velocities could result in false indications of high integrity char layers for intumescent open-state cavity barriers, or a false high degree of flame spread prevention. Due to the importance of air entrainment into the cavity, the burner should be positioned such

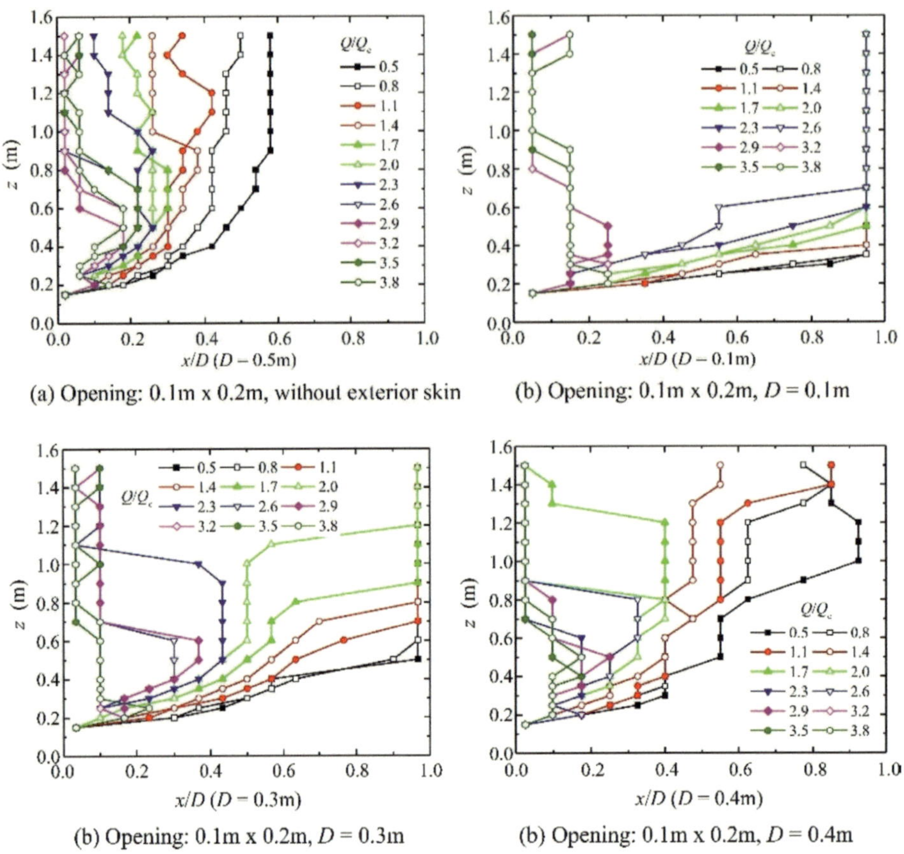

Figure 10.
Plume trajectory of a flame ejecting out of a window opening and into a cavity with different heat release rates and cavity widths. Reproduced from [34].

that the bottom of the cavity remains open. Neither TGD 19 nor prEN 1364_6 provide guidance on flow velocities. Although the ASTM approach does consider flow velocity, its assessment is limited to the influence of the cavity barrier's unexposed surface. It requires the velocity across the unexposed surface to be maintained below 1.3 m/s within the 3 minutes prior to ignition of the burner. Any future test method should consider monitoring the flow velocity within the cavity at the unexposed side of the cavity barrier as well as below the cavity barrier before and during testing. The velocity values should be compared with measured values from a cladding test to verify their appropriateness.

4.5 Instrumentation

The three test methods specify the use of thermocouples at different locations to measure the temperature development during the test, which is later used to assess against performance criteria. TGD 19 and prEN 1364_6 require thermocouples to be placed at every one-fourth of the total length of the specimen (**Figure 11a** and **b**, respectively). In TGD 19, three thermocouples are used at each location on the unexposed side of the cavity barrier. One is placed on the supporting wall, and the other two are on the unexposed surface of the cavity barrier and the top of the gap between the cavity barrier and the adjacent leaf. However, in prEN 1364_6, all the thermocouples are placed along the centre line of the cavity barrier, and careful usage of roving thermocouples has been enabled to measure temperature at any local "hot spots" and places where it is impractical to attach thermocouples. The ASTM E2912 approach also specifies the usage of unexposed side temperature measurements but has been limited to thermocouples attached to the vertical test assembly rather than the cavity barrier, **Figure 11c**. An improvement can be suggested regarding the usage of thermocouples at the exposed surface of the cavity barrier as well. Such monitoring

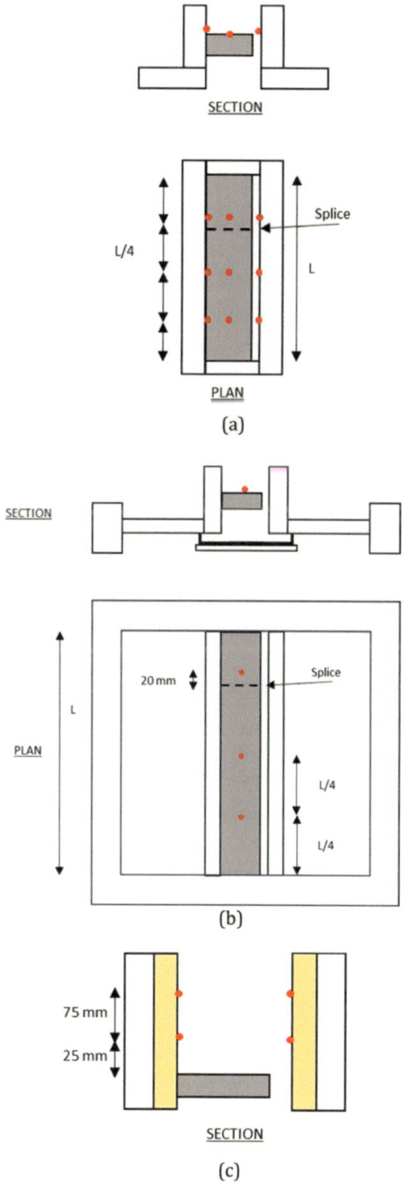

Figure 11.
Schematics of existing test methods used to evaluate the fire performance of open-state cavity barriers showing thermocouple locations for (a) TGD 19, (b) pr EN 1364-6 and (c) ASTM 2912.

is beneficial in verifying that the thermal conditions at the cavity barrier can replicate the fire conditions within a ventilated facade cavity.

4.6 Performance criteria

4.6.1 Existing performance criteria

The pass or failure of an open-state cavity barrier is decided based on the achievement of defined performance criteria. All three existing test methods include three performance criteria: closure time, insulation and integrity.

4.6.1.1 Closure time

The closure time criteria assess the performance of the open-state cavity barrier based on the time taken to close the cavity, usually with the expansion of the intumescent strip. The target values for this assessment have been specified as five minutes. According to prEN 1364_6 and ASTM 2912, it is only a visual measurement. TGD 19 addresses this criterion with more rigorous measurement by defining a "suspended" thermocouple temperature target of 180 K rise above initial ambient temperature coupled with the visual observation. The TGD 19 and prEN 1364_6 test methods specify the initial five minutes of the test to be considered as the time required to establish the specified temperature and pressure conditions. Thus, both test methods disregard the first five minutes for the closure time and the other performance criteria, discussed below.

Cavity fires can rapidly spread fire beyond the cavity barrier, especially when combustible materials are involved, it is therefore questionable whether the first five minutes of the test should be disregarded. A more appropriate measure could be to define the closure time target based on the combustibility class of the ventilated cavity. However, if the first five minutes are considered, additional performance targets should be defined for the temperature above the cavity barrier. Additionally, when the test incorporates a burner as the fire source, a maximum flame height beyond the cavity barrier before complete closure should be considered as another performance criterion to mitigate the risk of flame spread prior to the closure of the cavity barrier.

4.6.1.2 Insulation

The insulation criterion for assessing the open-state cavity barrier performance is similar for all existing test methods. To evaluate the insulation criteria, TGD 19 specifies a target value of 180 K above its initial temperature for the unexposed surface thermocouple temperature. The pr EN 1364_6 test method specifies this as an average unexposed side temperature increment of 140 K or any point unexposed side temperature increment of 180 K. ASTM E2912 also specifies average and maximum temperature targets as 139 K, ΔT and 181 K, respectively. As mentioned earlier, for the insulation criteria, the TGD 19 and prEN 1364_6 exclude failure within the first five minutes of the test, which can be crucial for cavity spaces, especially ones containing combustible materials. Therefore, for this criterion, the testing procedure should be consider scenarios where the first five minutes must be included, or specific target values need to be introduced to limit the fire spread during this period.

4.6.1.3 Integrity

For the integrity criteria, these three approaches are required to avoid sustained flaming at the unexposed side of the cavity barrier. To facilitate this, cotton pad testing methods have been suggested. The exclusion of failure within the first five minutes applies to the integrity criteria under TGD 19 and pr EN 1364_6 and still needs to be reconsidered.

4.6.2 Additional performance criteria

In addition to the existing performance criteria, the following criteria have been suggested for consideration in the development of future test methods.

4.6.2.1 Lateral spread

In TGD 19 and prEN 1364_6, the use of a furnace suggests the cavity barrier is exposed to approximately the same heating condition throughout its length. Thus, the cavity barrier seals the cavity throughout its length. However, regarding the fire penetration into a ventilated façade cavity, the fire exposure to the open-state cavity barrier is more likely a localised heat exposure. In this circumstance, the fire-exposed region of the cavity barrier will be activated to seal the gap. However, due to the low heat exposure, sealing may not be achieved further away from the localised fire penetration. Therefore, once the fire-exposed region is closed, smoke and flame can divert its path and pass the cavity barrier through the effectively unsealed areas. Thus, assessing the sealing capability of the open-state cavity barriers in the lateral direction is also essential. However, this could be done only using localised heating conditions, as in ASTM E2912. The suggested parallel panel test method can consider this performance assessment when a sufficient cavity length is used.

4.6.2.2 Smoke toxicity and spread

In addition to the existing performance criteria, some suggestions can be made. Combustible insulations and cladding materials in the ventilated facades can emit toxic gases, such as hydrogen cyanide and carbon monoxide during fire [9, 41]. When a cavity fire spreads, toxic smoke can enter the buildings through vent systems if the cavity barriers cannot seal the cavity and openings [9, 41]. Therefore, a specific performance assessment that the existing test methods should be newly introduced is the smoke spread and toxicity levels. For this, updated standards can measure the toxicity levels or smoke quantities before and after the closure of the open-state cavity barrier.

4.6.2.3 Durability

The discussed performance checks are conducted for samples in their pristine condition. However, during the operational life of a cavity barrier, it can undergo cyclic temperature variation and moisture ingression. These factors can play a crucial role in the durability of the cavity barrier, which can impact the fire performance. Therefore, new standards should also focus on assessing the durability of open-state cavity barriers and the fire performance under prolonged exposure to moisture and cyclic temperature.

5. Conclusions

The suitability of existing test methods to assess the fire performance of open-state cavity barriers used in ventilated facades is reviewed, and potential developments to future test methods are identified. TGD 19, prEN 1364_6 and ASTM E2912 are the only test methods currently being used to assess the fire performance of open-state cavity barriers. However, these test methods cannot effectively be used to determine the fire performance of open-state cavity barriers, and several modifications are suggested to address existing limitations. These test methods can be used to compare the different configurations and types of open-state cavity barriers. However, predicting the fire performance in the full-scale façade tests or real-scale façades is not possible due to the significant differences between the test setups and actual ventilated façade systems. These variations mainly concern the fire exposure conditions and cavity configurations.

The parallel panel test arrangement was identified as a potentially suitable test geometry to address the identified limitations. The existing performance criteria are suitable for use in future test methods. However, the exclusion of assessment within the first five minutes of fire exposure comes with risks associated with early fire spread and cavity barrier performance. Several suggestions have been made to improve the performance evaluation criteria. Evaluation of flame penetration heights, flow velocity, smoke toxicity and performance under prolonged exposure to cyclic moisture and temperature can further improve the application of these test methods. The suggestions and discussions are made in an effort to aid the development of future measures and policies on the fire safety of ventilated façade systems.

Conflict of interest

The authors declare no conflict of interest.

Author details

Kate T.Q. Nguyen*, Thomas W. Loh and Rajeendra Godakandage
School of Engineering, RMIT University, Melbourne, VIC, Australia

*Address all correspondence to: kate.nguyen@rmit.edu.au

References

[1] Baciu IR, Isopescu DN, Lupu ML, Maxineasa SG, Pruna L, Dan S. Ventilated façade solutions. IOP Conference Series: Materials Science and Engineering. 2022;**1242**:012002

[2] Rainscreen Facades Today. Fairview.com; 2019

[3] Newton E. How Do Ventilated Facades Improve Building Sustainability? Australia: Building Enclosure Online; 2023. Available from: https://fv.com.au/news/australia/rainscreen-facades-today

[4] Giraldo MP, Lacasta A, Avellaneda J, Burgos C. Computer-simulation study on fire behaviour in the ventilated cavity of ventilated façade systems. MATEC Web of Conferences: EDP Sciences. 2013;**9**:03002-p1 to 03002-p11. Available from: https://www.matec-conferences.org/articles/matecconf/abs/2013/07/matecconf_isfsf13_03002/matecconf_isfsf13_03002.html

[5] Asimakopoulou EK, Kolaitis DI, Founti MA. Performance of a ventilated-façade system under fire conditions: An experimental investigation. Fire and Materials. 2020;**44**:776-792

[6] Sharma A, Mishra KB. Experimental investigations on the influence of 'chimney-effect' on fire response of rainscreen façades in high-rise buildings. Journal of Building Engineering. 2021;**44**

[7] Livkiss K, Svensson S, Husted B, van Hees P. Flame heights and heat transfer in façade system ventilation cavities. Fire Technology. 2018;**54**:689-713

[8] Schabowicz K, Sulik P, Zawiślak Ł. Identification of the destruction model of ventilated facade under the influence of fire. Materials. 2020;**13**

[9] Peck G, Jones N, McKenna ST, Glockling JLD, Harbottle J, Stec AA, et al. Smoke toxicity of rainscreen façades. Journal of Hazardous Materials. 2021;**403**:123694, pp. 1-11. Available from: https://www.sciencedirect.com/science/article/pii/S0304389420316800

[10] Guillaume E, Dréan V, Girardin B, Benameur F, Koohkan M, Fateh T. Reconstruction of Grenfell tower fire. Part 3—Numerical simulation of the grenfell tower disaster: Contribution to the understanding of the fire propagation and behaviour during the vertical fire spread. Fire and Materials. 2020;**44**:35-57

[11] Čolić A, Pečur IB. Influence of horizontal and vertical barriers on fire development for ventilated façades. Fire Technology. 2020;**56**:1725-1754

[12] Guillaume E, Fateh T, Schillinger R, Chiva R, Ukleja S. Study of fire behaviour of facade mock-ups equipped with aluminium composite material-based claddings, using intermediate-scale test method. Fire and Materials. 2018;**42**:561-577

[13] Boström L, Skarin C, Duny M, McNamee R. Fire Test of Ventilated and Unventilated Wooden Façades. Sweden: SP Technical Research Institute of Sweden; 2016

[14] Jamison KLT, Boardman DA. A new fire performance test for cavity wall insulation. MATEC Web of Conferences. 2016;**46**:02004

[15] Bonner M, Wegrzynski W, Papis BK, Rein G. KRESNIK: A top-down, statistical approach to understand the fire performance of building facades using standard test data. Building and Environment. 2020;**169**:106540, pp. 1-10

[16] Nguyen KTQ, Weerasinghe P, Mendis P, Ngo T. Performance of modern building facades in fire: A comprehensive review. Electronic Journal of Structural Engineering. 2016;**16**:69-85

[17] Hassan MK, Hossain MD, Gilvonio M, Rahnamayiezekavat P, Douglas G, Pathirana S, et al. Numerical investigations on the influencing factors of rapid fire spread of flammable cladding in a high-rise building. Fire. 2022;**5**

[18] Godakandage R, Weerasinghe P, Gamage K, Adnan H, Nguyen K. A systematic review on cavity fires in buildings: Flame spread characteristics, fire risks, and safety measures. Fire. 2023;7:12

[19] Trafalgar F. Passive Fire Stopping for Cavity Barriers

[20] CFS-VB E60 Cavity Barrier 25mm Air Gap. Bahrain: HILTI; Available from: https://www.hilti.bh/c/CLS_FIRESTOP_PROTECTION_7131/CLS_PREFABRICATED_JOINT_7131/r21707845

[21] Siderise. Siderise RH Horizontal Cavity Barrier

[22] VFB plus. Tenmat

[23] Rakic J. Article: Cavity barriers and their application here in Australia

[24] Standardization Iof. ISO 13785-1:2002 Reaction-to-Fire Tests for Façades. Geneva, Switzerland: International Organization for Standardization; 2002

[25] Institution BS. BS 8414-2:2020: Fire Performance of External Cladding Systems—Test Method for Non-loadbearing External Cladding Systems Fixed to, and Supported by, a Structural Steel Frame. London: British Standards Institution; 2020

[26] Association TNFP. NFPA 285 Standard Fire Test Method for Evaluation of Fire Propagation Characteristics of Exterior Wall Assemblies Containing Combustible Components. USA: National Fire Protection Association (NFPA); 2023

[27] Institute SNTaR. SP FIRE 105,—Method for fire testing of façade materials

[28] (ASFP) AfSFP. ASFP Technical Guidance Document—TGD 19: Fire Resistance Test for 'Open-State' Cavity Barriers Used in the External Envelope or Fabric of Buildings; 2017

[29] Standardization Ecf. Draft BS EN 1364-6 Fire Resistance Tests for Non-loadbearing Elements. Part 6: Cavity Barriers. Brussels: European Committee for Standardization; 2022

[30] International A. ASTM E2912 Standard Test Method for Fire Test of Non-Mechanical Fire Dampers Used in Vented Construction. Pennsylvania, USA: ASTM International; 2017

[31] European Organization for technical A. EOTA Technical report TR 031: Fire resistance tests for cavity barriers; 2008

[32] American National Standard for Cleanroom Materials Flammability Test Protocol ANSI/FM Approvals 4910; 2004

[33] Global F. ANSI/FM Approvals 4411-2020—American National Standard for Cavity Wall Systems; 2020

[34] Miao L, Chow CL. A study on window plume from a room fire to the cavity of a double-skin façade. Applied Thermal Engineering. 2018;**129**:230-241

[35] Fogerty D. The Influence of Cavity Widths and Lateral Ventilation

Conditions on the Fire Behaviour within a Non-Combustible Ventilated Façade System. Lund, Sweden: Lund University; 2020

[36] Weerakkody N, White N, Moinuddin K. Evaluation of exterior wall cavity fires using an intermediate scale test method. Fire and Materials. n/a

[37] Hietaniemi J, Vaari J, Hakkarainen T, Huhta I, Jumppanen U-M, Korhonen T, et al. Fire Safety of Cavity Spaces—Characteristics of Fires in Building Voids, their Structural Prevention and Extinguishing. VTT Research Notes 2249. Espoo: VTT; 2004

[38] Jensen G. Fire spread modes and performance of fire stops in vented façade constructions—Overview and standardization of test methods. MATEC Web of Conferences: EDP Sciences. 2013;**9**:02002-p.1 to 02002-p.11. Available from: https://www.matecconferences.org/articles/matecconf/abs/2013/07/matecconf_isfsf13_02002/matecconf_isfsf13_02002.html

[39] Tyrol M. FM4411—Approval standard for cavity walls and rainscreens. In: Air Barrier Association of America, Conference and Trade Show; 8-9 May 2018; Salt Lake City, Utah. 2018. Available from: https://www.abaaconference.com/wp-content/uploads/2018/05/Mark-Tyrol-FM4411-%E2%80%90-Approval-Standard-For-Cavity-Walls-And-Rainscreens.pdf

[40] Approvals F. Cavity Wall Standard Being Revised to Reduce Risk

[41] Chen TBY, Yuen ACY, Yeoh GH, Yang W, Chan QN. Fire risk assessment of combustible exterior cladding using a collective numerical database. Fire. 2019;**2**:1-14

Chapter 4

Inorganic Polymers (Geopolymers) Based on Construction and Demolition Waste for Fire Safety of Buildings and Constructions

Ioanna Giannopoulou and Demetris Nicolaides

Abstract

Fire safety of buildings and civil engineering structures is of paramount importance for people's safety and the protection of their property. In this paper, the development of a new, sustainable and advanced fire-resistant inorganic polymer (geopolymer) is investigated. The new material is based on the recycling of construction and demolition waste, especially the waste brick stream. The behavior of the developed inorganic polymer at elevated temperatures was evaluated in terms of physical, thermal and mechanical properties. The waste brick-based inorganic polymer was thermally stable up to 1050°C with residual compressive strength ranging from 15 to 25 MPa. The compression stress under thermal load revealed elastic deformation before fracture up to 1000°C, and plastic deformation without fracture at higher temperatures. The strength variations observed at elevated temperatures were attributed to the mineralogical transformations occurring in the geopolymeric matrix. The fire performance of the inorganic polymer was also tested according to the ISO-834 standard time–temperature curve. The material managed to keep the temperature at the interface with concrete below the temperature set by the ISO-834 fire test (180°C) for 100 min, proving its effectiveness to be used as a thermal barrier in buildings and constructions. The specimen exposed to fire kept its structural integrity after the test completion, without presenting any significant macroscopic damage. Based on the results of this research, the developed inorganic polymer is a suitable material to be used for the fire safety of buildings and other construction projects.

Keywords: inorganic polymers, geopolymers, fire resistance, passive fire protection, building, constructions, sustainability, recycling

1. Introduction

Over the past 50 years, serious fire incidents that have occurred worldwide in residential and high-rise buildings and other constructions, such as road and railway tunnels, caused numerous human deaths and injuries, as well as costly structural damage with significant consequences for the regional economy and society. Building and construction materials, such as concrete, bricks, plaster and steel, are heavily damaged when exposed to the high temperatures developed in fire, jeopardizing the integrity and stability of structures. During a fire, concrete is subjected to severe spalling and cracking, which may result in the collapse of the structure. Concrete spalling is strongly related to the huge amount of vapor released by the temperature of 380°C [1], due to the dehydration of calcium hydrates. This phenomenon generally results in the destruction of concrete integrity and the reduction of its mechanical strength, leading to the detachment of large parts from structural components. Similarly, bricks and plasters lose their coherence and deform or become brittle in fire, while the steel expands and softens, losing its mechanical strength and stiffness [1], which may lead to excessive or permanent deformation. All these phenomena reduce the load-bearing capacity of structures exposed to fire, usually rendering them inappropriate and dangerous for use, and in the worst case, leading to their collapse. In certain cases, neighboring structures are also greatly affected by building fires. To minimize the devastating effects of a fire and, above all, to protect human lives, the passive fire protection of buildings and constructions is essential, while it is also mandatory, according to the new building codes worldwide [2]. Unlike the active fire protection systems of structures (sprinklers, fire extinguishers, etc.) that require human intervention or mechanical activation to operate, the passive fire protection systems rely on building-incorporated materials that act as natural barriers against fire. These materials seek to preserve the integrity of buildings and constructions under fire, for a sufficient time to enable safe evacuation of people and limit the structural damage.

Today, fireproof plasters and boards of gypsum and calcium silicate are mainly used to protect the concrete and steel structural elements against fire [3, 4], while new groups of materials like the spray fire-resistant aggregates of expanded perlite and vermiculate and the paint-like intumescent coatings [5–7] have been also developed to act as firewalls in buildings and constructions. However, important drawbacks of these materials, like the elevated cost, the long installation time and the inability to withstand temperatures higher than 750°C [8], inhibit their wide application as fire protective building solutions. There is, therefore, an open field for low-cost and highly effective fire-resistant materials that would cope with building fires, meeting the strict fire safety requirements for buildings and constructions. In this context, inorganic polymers (geopolymers) are advantageous materials, as they combine low production costs with excellent fire performance. These materials are based on the geopolymerization process, which involves a sequence of complex chemical reactions between solids rich in reactive aluminosilicate phases and alkali-silicate solutions, under high alkaline conditions. Geopolymerization yields the formation of amorphous to semi-crystalline three-dimensional frameworks consisting of chain and ring polymers with Si and Al atoms in IV-fold coordination with oxygen [9]. Positive ions such as Na^+, K^+ and Ca^{2+} are also present in this framework to balance the negative charge of Al^{4+} [9]. The processing temperature of geopolymerization is low, below 100°C, resulting in materials with extremely low embodied energy and carbon footprint compared to the traditional ceramic building materials and concrete [10]. In terms of physical, mechanical and thermal properties, inorganic polymers are comparable to or even better than conventional building and construction materials. They present low apparent porosity, negligible shrinkage, high mechanical strength and acid resistance, while they are heat- and fire-resistant materials, inflammable and incombustible, and do not release hazardous and toxic gases, even after extended exposure to high temperatures. Given that a variety of industrial waste and residues can be utilized for their production, inorganic polymers are considered sustainable alternatives to several building and construction materials, with great potential in applications that require combined mechanical and thermal properties.

The unique fire performance of inorganic polymers (geopolymers) is attributed to the inherent fire resistance they exhibit and is related to their chemistry and microstructure. In a fire, the absorbed and chemically bonded water molecules and hydroxyl ions existing in the mass of these materials evaporate through endothermic reactions (dehydration and dehydroxylation), which absorb heat, thus reducing the surrounding thermal load [11, 12]. In addition, the micro- to nano- porosity existing in inorganic polymers allows the water to evaporate without increasing the pressure of internal porosity and damaging the materials microstructure [11–13]. The development of fire-resistant geopolymers was studied for the first time by J. Davidovits in 1972, after the catastrophic tunnel fires in France [14]. Since then, several research studies have been published regarding the advancements of fire-resistant inorganic polymers.

The aluminosilicate raw materials used in these studies include natural clays like metakaolin and industrial wastes, such as ashes, metallurgical slags, mine tailings, waste glass and construction and demolition waste. Barbosa and McKenzie [15] investigated the thermal behavior of metakaolinite-based inorganic polymers with SiO_2/Al_2O_3 and Na_2O/SiO_2 ratios equal to 3.3 and 0.25, respectively. The derived materials exhibited remarkable thermal stability up to around 1300°C, before starting to melt. Besides, the composites of these inorganic polymers with granular inorganic fillers exhibited good thermal performance up to 1000°C, with little detectable reaction between the filler and the geopolymer matrix. Different mixtures of fly ash and metakaolin were also used to develop geopolymers with thermal stability at high temperatures [16]. However, the mechanical strength of the obtained materials was constantly decreased with temperature increasing, while several cracks appeared on their surfaces from the temperature of 300°C. Similarly, the class F fly ash-based geopolymers prepared by using both, sodium- and potassium- based alkali activators proved unsuitable for refractory applications, as they showed high shrinkage and low thermal stability when exposed to temperatures ranging from 800 to 1200°C [17]. Moreover, the residual compressive strength of geopolymers based on combinations of fly ash and slag in different ratios decreased exponentially with the increase of their initial strength, after their exposure to high temperatures, up to 800°C [18]. This behavior was attributed to the decrease in specimens' ductility that was observed with the increase of the initial compressive strength. On the contrary, the compressive strength of inorganic polymers based on granulated blast furnace slag (BFS) rich in calcium was constantly increased upon their exposure to temperatures from 200 to 1000°C for 1 hour [19], almost doubling the initial value. Furthermore, the reverse-side temperature of BFS-based alkali-activated materials exposed to 1100°C flame was equilibrated at less than 400°C after 35 min [20]. In addition, the fire performance of inorganic polymers based on a ferronickel slag was successfully assessed according to the ISO-834 and rijks water staat (RWS) standard temperature–time curves [21]. The ISO-834 standard fire curve was also used to investigate the fire performance of foamed and unfoamed metakaolin-based geopolymers reinforced with polypropylene fibers [22] to assess their suitability for thermal barriers and fire-resistant panels. Based on the results, the unfoamed geopolymer presented better performance than the foamed one. Finally, inorganic polymers prepared by specific streams of Construction and Demolition Waste (CDW) exhibited very good thermal and mechanical properties after exposure to elevated temperatures up to 1050°C, for 2 hours [23].

Among the above-mentioned industrial wastes, CDW can be found everywhere and therefore, is the most suitable to meet the challenges of the circular economy in the building and construction industry. This waste results from the demolition, the renovation, and the construction of buildings, roads, bridges, tunnels, etc., and comprises a wide group of waste materials, including concrete, bricks, ceramics, gypsum, wood, glass, metals, plastic, organic materials and excavated soil. In the European Union (EU), around 450–550 million tons of CDW is produced every year, accounting for more than 35% of the total industrial waste volume generated per annum [24]. Similar quantities of CDW are generated every year in the United States (USA), of which only about 5% comes from the residues of the construction industry, with the remaining 95% coming from demolition activities [25]. Approximately 12% of CDW is concrete, while more than 54% corresponds to ceramic-type materials, like bricks and ceramic tiles. According to previous research, the recycling of CDW in the production of inorganic polymers intended for the building and construction industry [26–30] is an attractive and effective option for the sustainable management of this waste.

In this paper, the development of a new, sustainable fire-resistant inorganic polymer (geopolymer) based on waste bricks of the CDW is investigated. The obtained material was evaluated in terms of mechanical properties and thermal stability, after exposure to temperatures up to 1050°C for 2 hours. The mineralogical transformations that occurred in the geopolymeric matrix at elevated temperatures were investigated by the XRD method. Furthermore, the compression deformation of the developed material subjected to thermal load, as well as its fire performance according to the ISO-834 standard time–temperature curve, were assessed. According to the findings of this research work, the recycling of waste bricks for the development of fire-resistant inorganic polymers intended for the passive fire protection of buildings and construction is very promising.

2. Experimental procedure

2.1 Raw materials

Waste Bricks (WB), which is a stream of Construction and Demolition Waste (CDW), was used in this research as a solid precursor source of inorganic polymers. An adequate quantity of WB for the development of fire-resistant inorganic polymers was delivered from a CDW recycling plant in Cyprus. Waste bricks were crushed and milled to a particle size below 1.18 mm. The chemical composition of WB was determined by the energy-dispersive X-ray fluorescence method (ED-XRF; SPECTRO XEPOS) and presented in **Table 1**. As shown in **Table 1**, WB was rich in silicon and aluminum oxides, indicating suitability for geopolymerization, while it also contained important quantities of iron and calcium oxides, as well as small amounts of magnesium and potassium oxides.

The mineralogical analysis performed on a D8 Bruker Diffractometer (CuKα radiation, $\lambda = 1.5418$ Å, tungsten filament) detected quartz (SiO_2) and the feldspars anorthite ($CaAl_2Si_2O_8$) and albite ($NaAlSi_3O_8$) as predominant phases in this waste material. Moreover, minor amounts of calcite ($CaCO_3$) and hematite (Fe_2O_3) were present, while diopside ($MgO{\cdot}CaO{\cdot}2SiO_2$), wuestite ($FeO$) and mullite ($3Al_2O_3{\cdot}2SiO_2$) were tentatively identified.

Oxide	Mass (wt. %)
SiO_2	53.57
Al_2O_3	14.33
CaO	7.71
FeO	10.19
K_2O	3.74
MgO	4.07
Na_2O	0.66
TiO_2	1.46
Other	4.27

Table 1.
Chemical composition of WB.

According to the particle size distribution analysis of WB, which was performed using a Mastersizer 500 laser analyzer (Malvern Instruments Ltd.), the mean particle size was 43 μm. In this experimental work, only the −250 μm class of WB was used, as received after sieving. The used class of WB had a bulk density of 1.63 g/cm^3, an apparent density of 2.81 g/cm^3 and water absorption equal to 25.57%.

A hybrid alkaline silicate activator, which combined a potassium hydroxide (KOH) solution of 8 M (mol/L) concentration and a sodium silicate solution (MERCK S.A., Na_2O = 8%, SiO_2 = 27%, H_2O = 65% and d = 1.346 g/mL), was used to drive the geopolymerization reaction in this work. The potassium hydroxide solution was prepared by dissolving KOH anhydrous pellets (MERCK, 99.5% purity) in distilled water. After its preparation, the KOH solution was left for at least 24 hours before being used. The two solutions were mixed and stirred on a stirring plate for 5 min before their use.

2.2 Synthesis of materials

The synthesis of the inorganic polymer studied in this paper is based on previous research [23]. The fire-resistant inorganic polymer based on the waste brick (BFR) was synthesized by using 100% WB and a hybrid alkali-metal silicate solution of a silicate modulus (M_s) equal to 1.1. The solid-to-liquid ratio (S/L) was 2.5 g/L. The two components were mixed in a planetary mixer (Hobart) for 5 min at a speed of 95–110 rpm to form a homogeneous paste. The paste was poured into steel cubic molds of 50x50x50 mm^3 and sealed with a plastic film to prevent water evaporation. The samples were cured at a temperature of 50°C for 7 days, under ambient pressure and uncontrolled humidity conditions. After curing, the specimens were demolded and left to harden under ambient conditions for 7 and 28 days, before undergoing the different tests and measurements. Two prismatic specimens of BFR with dimensions 100x100x40 mm^3 and 400x400x40 mm^3 were also prepared for use in the thermal conductivity measurement and fire resistance test, respectively.

2.3 Analysis and testing methods

The mineralogical analysis of the raw material and the inorganic polymers was performed according to the X-Ray Diffraction (XRD) method, using a D8 Bruker

Focus diffractometer (CuKα radiation) in the range of 2-theta from 2° to 70° with a scanning speed of 2°/min.

The microstructure of inorganic polymers was evaluated by the Fourier Transformation Infra-Red (FTIR) method in a Perkin Elmer 2000 spectrometer, using the Attenuated Total Reflection (ATR) technique with a Zn/Se crystal. FTIR spectra were collected in the wave number range from 650 to 4000 cm^{-1}.

The thermal conductivity of the BFR inorganic polymer was measured according to the guarded hot plate method on a Netzsch HFM 446 Lambda Series—Heat Flow Meter, using prismatic samples measuring 100x100x40 mm^3. The reported value is the average of five measurements.

The uniaxial compression strength of the developed materials was assessed using a universal mechanical machine with a 2000 kN measuring range. The prepared inorganic polymers were tested 7 and 28 days after curing, while those exposed to elevated temperatures were tested once they cooled down to room temperature. Three cubic specimens of 50 mm edge were used in each test, and the reported value corresponds to the average of the relevant measurements. The same cubic specimens were also used to define the density and mass loss of materials, and the reported values are the average of three measurements.

The thermal and mechanical behavior of the BFR inorganic polymer at elevated temperatures was assessed using two different procedures. The first procedure investigated the compressive strength of BFR after exposure to selected temperatures, while the second one focused on its compression deformation under thermal load. In the first procedure, three cubic specimens of 50x50x50 mm^3 were heated to the temperatures of 600, 800 and 1050°C in a laboratory electric furnace (**Figure 1a**). The furnace was heated at a rate of 4.4°C/min up to the desired temperature and left at this temperature for 2 hours, simulating the duration of a building fire. After this time, the samples were removed from the oven and left to cool back to ambient temperature in open air conditions. After the test, the exposed specimens were thoroughly inspected for cracks and surface alterations and tested for compressive strength, density and mass loss. Moreover, the rearrangements that occurred in the matrix BFR inorganic

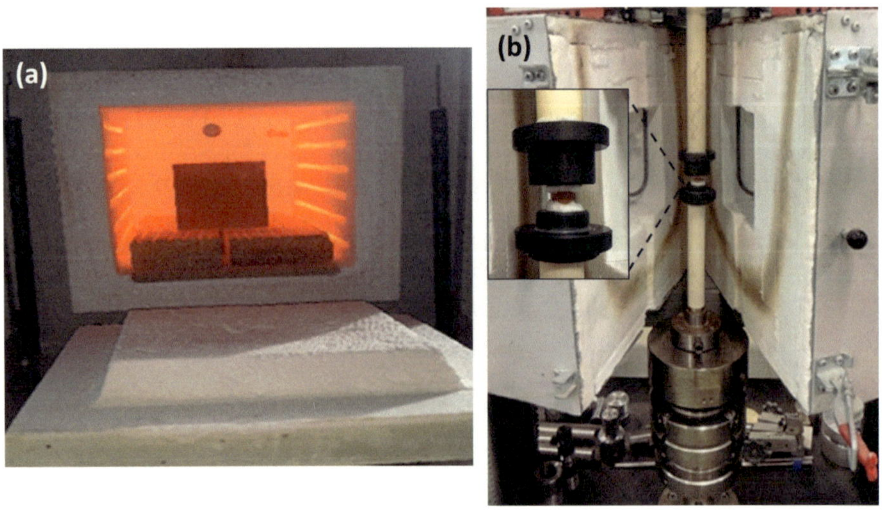

Figure 1.
(a) Electric furnace and (b) Instron 5982 loading machine with furnace (inset photo).

polymer were evaluated by XRD and FTIR analyses. In the second testing procedure, a universal mechanical testing machine, Instron 5982 (100 kN) equipped with a furnace of high temperatures (**Figure 1b**) with the possibility to reach up to 1300°C, was used to assess the compression stress of the BFR inorganic polymer under heat flux that simulated fire conditions. In this test, three cylindrical samples of 12 mm diameter and 16 mm height were heated to the temperatures of 1050, 1000 and 950°C, at which they were compressed between two alumina rods fitted with silicon carbide. To protect the furnace in case of a sample melting, alumina/silica foam and steel slims were used at temperatures below and above 1000°C, respectively, as insulation between the sample and the rods. In all tests, a heating rate of approximately 5°C/min and a displacement rate of 500 mm/min were applied, while temperature accuracy was 5°C.

The fire resistance of the BFR inorganic polymer was assessed according to the ISO-834 standard time–temperature curve (Fire resistance tests—Elements of building constructions) [31]. This standard test is based on the burning rate of commonly used materials in the building and construction industry (fire of cellulosic nature) and defines a testing method for evaluating the fire resistance of materials and systems intended for the passive fire protection of building and construction components. In this study, a prismatic specimen of BFR inorganic polymer (fire-resistant material) with dimensions of 400x400x40 mm^3 was fixed by appropriate fire-resistant anchors (Fischer FNA 6x30/30 A4) on a 400x400x200 mm^3 concrete slab (fire-protected component) that was prepared according to European federation of national associations representing producers and applicators of specialist building products for concrete (EFNAR) specifications [32]. The resulting combined specimen was then placed at the opening of a custom-made electric furnace, so that the inorganic polymer to be directly exposed to the heat flux inside the furnace, as shown in **Figure 2a**. Eleven thermocouples of K type (NiCr-Ni), which were placed at specific points of the specimen (**Figure 2b**), monitored and continuously measured the temperature at the interface of BFR with the concrete slab, as well as into the concrete slab, at the distance of 25 mm, 40 mm and 75 mm from the interface and on the back side.

Two more thermocouples were used to monitor the temperature inside the furnace. The temperature in the furnace was increased following the temperature evolution of the ISO-834 standard curve, as is defined by Eq. (1).

$$T = \mathrm{T}_0 + 345 \cdot \log_{10}(8 \cdot t + 1) \tag{1}$$

where T is the furnace temperature (°C), t is the time (s) and T_0 is the initial furnace temperature (°C).

3. Results and discussion

3.1 Physical mechanical and thermal properties

The compressive strength of the BFR inorganic polymer was 19.2 MPa after the first 7 days of hardening, while it reached 22.7 MPa after 28 days of hardening, in total. The density of BFR was 1556 kg/m^3 and remained almost unchanged throughout the whole hardening period. Moreover, this material presented negligible shrinkage (< 1%) and weight loss (< 3%) in the same period. Finally, the thermal conductivity of the BFR inorganic polymer was detected to be 0.156 W/ (m·K) at 25°C. Such a thermal conductivity value is similar to or even better than that of commercial

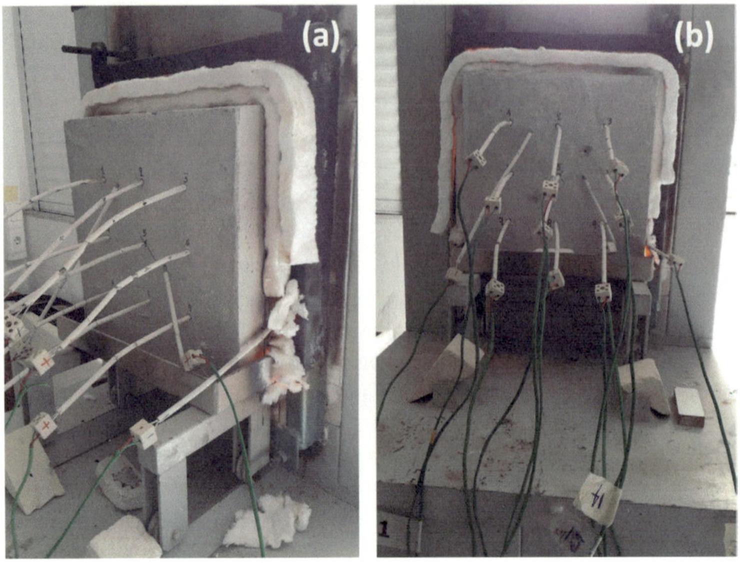

Figure 2.
(a) Experimental set-up of the fire resistance test and (b) arrangement of thermocouples.

fire-resistant materials currently used in the construction industry [33], rendering the BFR inorganic polymer a promising candidate for passive fire protection applications.

3.2 Elevated temperature mechanical behavior

In **Figure 3**, the compressive strength of the BFR inorganic polymer exposed to elevated temperature is plotted as a function of hardening time (7 and 28 days) and temperature of exposure. For comparison purposes, the compressive strength of the BFR, after curing (50°C), is also included in **Figure 3**.

As shown in **Figure 3**, the compressive strength of BFR followed a similar trend for both hardening periods. As regards the materials that had undergone hardening for 7 days, compressive strength decreased to 15.4 MPa after their exposure to 600°C. Then, the compressive strength increased with temperature, reaching its initial value (19.4 MPa) after exposure to 800°C, which is slightly increased (21.06 MPa) to the highest studied temperature of 1050°C. Similarly, the compressive strength of BFR specimens that were left 28 days for hardening decreased to 18.8 MPa when they were exposed to 600°C and then it increased to 23.7 and 25.2 MPa, after exposure to 800 and 1050°C, respectively.

The decrease in compressive strength observed after the exposure of BFR to 600°C (**Figure 3**) can be attributed to the massive evaporation of the chemically bonded water from the geopolymeric matrix of the material. At this temperature, the water molecules and hydroxyl ions that participate in the hydrate and hydroxylate alkali-silicate-aluminate phases are removed from the geopolymeric matrix, creating an increased vapor pressure in the micro- to nano-porosity of these materials that negatively affects their compressive strength [23]. In addition, at around 500–550°C, the softening of the alkali-aluminosilicate gel phases starts to occur [34], contributing to the decreased compressive strength of materials. The softening process of the geo-polymeric gel matrix is also responsible for the improvements in compressive strength observed at higher temperatures (**Figure 3**). The viscous phase that is formed involves

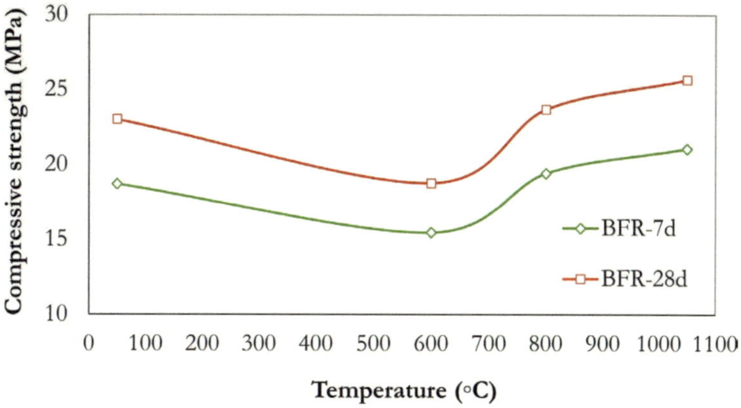

Figure 3.
Compressive strength of the BFR inorganic polymer after exposure to elevated temperatures for 2 hours (curing at 50°C for 7 days; hardening at ambient temperature for 7 and 28 days).

critical changes in the structure of inorganic polymers (geopolymers) with the increase of temperature. This phase fills the internal porosity of materials and sinters the particles of the solid precursor, leading to the densification of materials to glass- or ceramic-like structures [35]. Moreover, at elevated temperatures, it also promotes the cracks healing, thus improving the durability of materials.

The mineralogical transformation that occurred in the BFR inorganic polymer after its exposure to elevated temperatures (**Figure 4**) justified the above-mentioned microstructural alterations.

As seen in **Figure 4**, the only mineralogical alteration identified in the BFR inorganic polymer after its exposure to 600°C (**Figure 4b**), is related to the kalsilite ($KAlSiO_4$) phase (**Figure 4a**). With the increase of the exposure temperature to 800°C, some important mineralogical changed were detected, including the

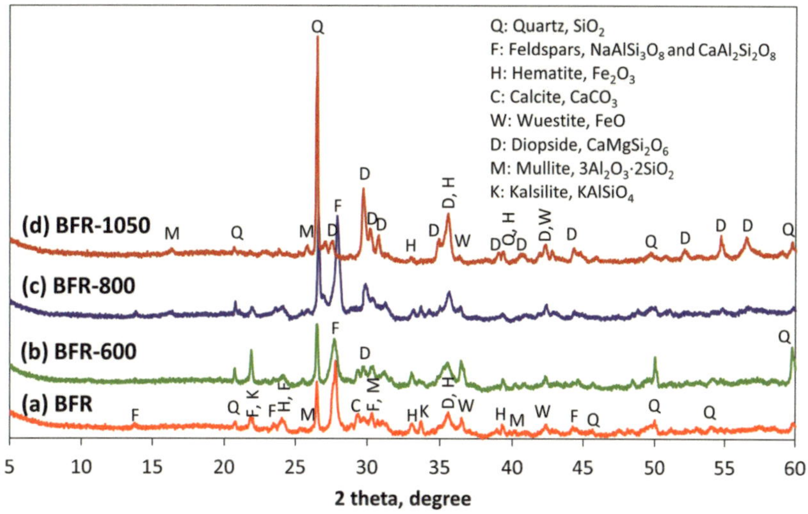

Figure 4.
XRD patterns of BFR inorganic polymer after (a) curing (50°C, 7 days) and (b), (c) and (d) exposure to 600, 800 and 1050°C, respectively, for 2 hours.

disappearance of the slight broad halo peak registered in 2-theta between 14° and 21° at the XRD pattern of the as-cured BFR (**Figure 4a**), which was associated to the amorphous gel phase formed after the geopolymerization of waste brick, as well as the decrease of quartz and the increase of diopside peaks intensity (**Figure 4c**). At the highest investigated temperature, more significant mineralogical changes were observed (**Figure 4d**). Feldspars that dominated in the BFR up to the temperature of 800°C (**Figure 4a–c**) disappeared after exposure to 1050°C, giving way to the crystallization of diopside and quartz phases, which both increased significantly in intensity and became predominant. The disappearance of the amorphous geopolymeric gel phase and the enhancement of diopside and quartz crystalline phases, which present increased hardness than feldspars, favored the mechanical strength of BFR (**Figure 3**).

The changes in the BFR density after its exposure to elevated temperatures were insignificant. The density was decreased from 1556 kg/m^3 to 1475 kg/m^3 at 600°C and remained quite constant up to 1050°C (1436 kg/m^3). The inorganic polymer also showed negligible mass loss (< 6%) and shrinkage (< 5%), upon its exposure up to 1050°C. Besides, it remained structurally intact to all testing temperatures, with no signs of spalling, deformation or failure, except for a few small cracks that were visible on the specimens' surface, after their exposure to 600°C. It is worth noting that these cracks disappeared after exposure to the temperatures of 800°C and above, due to the above-mentioned sintering phenomena that took place at these temperatures.

3.3 Compression deformation under thermal load

Figure 5 presents the stress–strain curves of the BFR inorganic polymer obtained by the gradual applying of load at the temperatures of 950 and 1000°C. Generally, the stress–strain curve quantifies the relationship between the load applied to the cross-sectional area of a test specimen and the resulting displacement and is a measure of the occurred deformation. The curve of the BFR specimen tested at the temperature of 1050°C is not included in **Figure 5**, since it underwent plastic deformation from the beginning of the test. The specimen was deformed without obvious fracture up to approximately 50% apparent strain, eventually taking on a permanent barrel-like shape (**Figure 6d**). The plastic deformation at this temperature is reasonable, as it is related to the softening of sodium silicates (melting point ~1080°C) in the

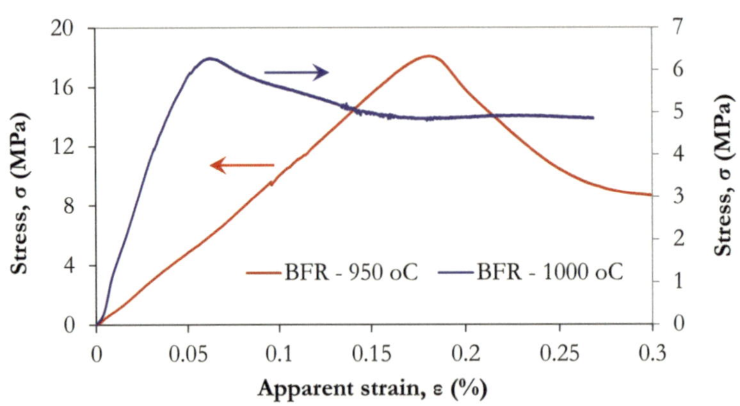

Figure 5.
Stress–strain curves of the BFR inorganic polymer at 950 and 1000°C.

Figure 6.
BFR specimens after stress–strain testing at (a) 50°C (b) 950°C, (c) 1000°C and (d) 1050°C.

geopolymeric matrix and the formation of a pseudo-viscous phase that negatively affects the cohesion of the materials [36].

According to **Figure 5**, the stress–strain curves of BFR specimens at both temperatures of 950 and 1000°C present a linear first stage, that indicates elastic deformation for the material in both temperatures. This behavior characterizes brittle materials like hard ceramics. In the linear elastic stage, stress is proportional to strain, with the slope indicating the modulus of elasticity. As shown in **Figure 5**, the slope is higher at 1000°C than at 950°C, demonstrating a higher ability of BFR to resist deformation at 950°C. Indeed, BFR underwent elastic deformation up to 0.2% strain at 950°C, which ended up with the specimens' fracture, while the elastic deformation of BFR was limited to a strain of up to 0.05% at 1000°C (**Figure 5**) followed by a small region of plastic stress.

As appeared in **Figure 6b** and **c**, BFR specimens presented a brittle fracture with low deformation at both temperatures of 950 and 1000°C. A strain hardening region is generally absent (**Figure 5**); thus, energy is absorbed by fracture, while the cracks formed under the applied load were unstable (**Figure 6b** and **c**), meaning that they could continue growing spontaneously, without the load increasing. The yield strength of BFR at 950°C was 18 MPa (**Figure 5**), which was similar to the residual compressive strength after 2 hours exposure to the same temperature. This means that the BFR inorganic polymer can withstand a load of about 18 MPa up to 950°C and recover its original shape when this load stops applying. On the contrary, the yield stress of BFR at the temperature of 1000°C was around 6 MPa (**Figure 5**), which was approximately one-third of the compressive strength developed after 2 hours exposure at this temperature. As seen in **Figure 6c**, many heavy cracks were formed on the BFR specimen, strongly affecting its structural integrity.

3.4 Fire resistance test performance

The experimental results of the BFR testing under the ISO-834 standard time–temperature curve are presented in **Figure 7**. In the corresponding graph, the temperature that was recorded at the interface between the concrete and the inorganic polymer, as well as inside the concrete slab at 40 mm distances from the interface and on the unexposed surface (back side) of the concrete slab, is plotted versus time. Moreover, the change in furnace temperature over time and the ISO-834 standard fire curve, as it was determined according to Eq. (1), are also included in **Figure 7**.

The performance criteria of a fire-resistant material that is tested according to the ISO-834 standard fire curve require: (i) the average temperature at the unexposed side of the specimen not to exceed the initial temperature by 155°C (this means that the temperature should not exceed 180°C, given an initial temperature of 25°C) and (ii) any crack appearing after the test completion on specimen's exposed surface should be lower than or equal to 6 mm thick for a length of 150 mm or more and not have any opening larger than 25 mm.

As shown in **Figure 7**, the temperature at the interface between the concrete and the BFR inorganic polymer failed to remain below 180°C throughout the test, as required by the ISO-834 standard fire curve. Specifically, this temperature slightly exceeded 180°C after 100 min, reaching the temperature of 224°C after 120 min. According to the experimental results given in **Figure 7**, in the first 10 min, when the furnace temperature was rapidly increased to 670°C, the temperature developed at the concrete-BFR interface was very low, about 18°C. In the following 50 min (60 min from test starting), the temperature at the concrete-BFR interface increased to 87°C, with the furnace temperature being 960°C, establishing a temperature gradient across BFR of 21.8°C/mm. Then, the temperature at the interface continued increasing at a rate of 2.3°C/min and reached 224°C at the end of the test, when the furnace temperature was 1074°C (**Figure 7**). At this time, the temperature gradient determined across BFR was almost constant and equal to 21.3°C/mm. Regarding the temperature developed inside the concrete slab at a distance of 40 mm from the concrete-BFR interface, it remained below 180°C until the end of the test. Finally, the temperature at the unexposed side of the concrete slab was less than 70°C during the fire test (**Figure 7**). At the end of the test, the BFR specimen presented coherence and integrity (**Figure 8**), without observing any structural damage, deformation, or important surface cracking on both sides, the one exposed at the thermal load (**Figure 8c**) and the other that was laid on the concrete slab (**Figure 8b**). Only some small-sized cracks have appeared on the exposed surface (**Figure 8c**), which did not affect the mechanical stability of the material. Moreover, the BFR inorganic polymer was proven non-combustible material.

The change of the BFR specimen color from red (**Figure 8a**) to brownish red (**Figure 8c**) observed before and after being subjected to the ISO-834 standard fire test, was associated with the transformation of the mineralogical phases that occurred at the developed temperatures. The high concentration of iron oxides in the BW

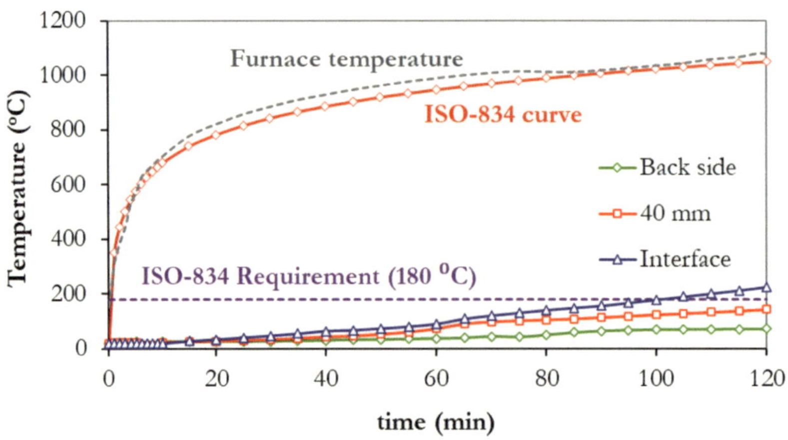

Figure 7.
Performance of the BFR inorganic polymer under the ISO-834 standard fire curve.

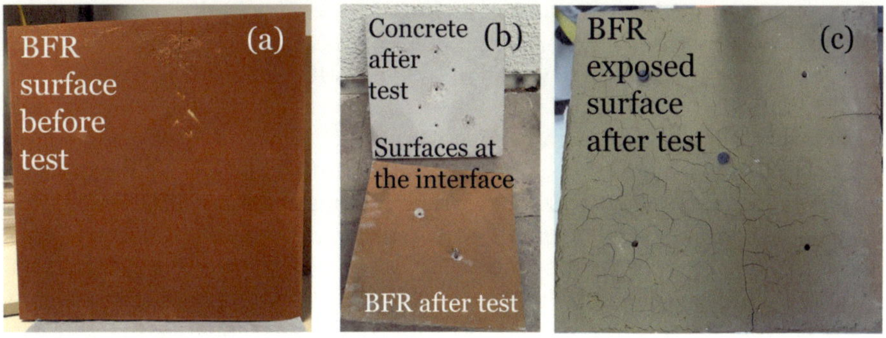

Figure 8.
The BFR specimen (a) before and (b) and (c) after the fire test.

(**Table 1**) in the form of hematite and wuestite (**Figure 4a**) was responsible for the red color of the BFR specimens before testing. During the fire test, the formation of diopside at around 900°C (**Figure 4d**) provided with a yellowish color the BFR specimens, resulting in the final brownish red color [37].

Although the BFR inorganic polymer failed to meet the temperature-related performance criterion required by the ISO-834 standard fire curve, it succeeded in keeping the temperature at the interface between concrete and geopolymer below 180°C for more than 100 min, which is very important for the safe evacuation of a building in case of fire. Overall, the BFR inorganic polymer succeeded in maintaining the temperature at the interface with concrete below 250°C after 2 hours of fire exposure, which is significant for the structural integrity of concrete, as this temperature is much lower than the critical temperature of the concrete exfoliation and spalling (380°C).

4. Conclusions

In the present work, the utilization of waste bricks derived from CDW, for the development of an advanced and sustainable inorganic polymer (geopolymer) for the fire safety of buildings and constructions was investigated. The most important findings of this investigation are the following:

The developed inorganic polymer showed compressive strength of 19.2 and 22.7 MPa withing 7 and 20 days, respectively, after curing. Its thermal conductivity was 0.156 W/ (m·K) at 25°C and its density was equal to 1556 kg/m^3. The physical, thermal and mechanical properties of the inorganic polymer were comparable to those of commercial fireproofing materials.

After exposure to 600°C for 2 hours, the compressive strength decreased by 17.5 to 18.5%, regardless of the hardening time. The compressive strength increased with further increase in exposure temperature and reached the initial value at 800°C, while it exceeded it by approximately 10% at the highest studied temperature of 1050°C. These alterations in compressive strength were attributed to the mineralogical transformations that occurred in the geopolymeric matrix. The inorganic polymer showed negligible shrinkage and mass loss and maintained its structural integrity at all exposure temperatures.

The compression stress of the inorganic polymer under heat load revealed elastic deformation up to 950°C, elastic to plastic deformation between 950 and 1000°C, and plastic deformation without fracture, at 1050°C. Up to 1000°C, the inorganic polymer presented brittle fracture, which is common for hard ceramic materials. The developed material was proved able to withstand a load of 18 MPa and 6 MPa under heat flux at 950 and 1000°C, respectively, before failure.

The performance of the inorganic polymer under the ISO-834 thermal load was very satisfactory. The material succeeded in the test for 100 min, proving that it has the ability to operate as a heat flux barrier for buildings and constructions. No significant cracks or any other deformation and defects were observed on both sides of the tested specimen, after 2 hours of exposure to the high thermal load. Some small-size cracks, which appeared on the exposed surface of the specimen, were deemed insignificant for affecting its structural stability. Besides, no concrete spalling was observed during and after the fire testing.

Overall, the good thermal and mechanical performance of the developed inorganic polymer (geopolymer) in simulating fire conditions, in combination with the utilization of waste for their production, the remarkably low processing temperature, and reduced carbon footprint, makes it attractive as fire-resistant material. However, a more detailed and comprehensive study of the long-term performance characteristics and the up scaling of the new material is required to better define its exact applications.

Conflict of interest

The authors declare no conflict of interest.

Author details

Ioanna Giannopoulou[1*] and Demetris Nicolaides[1,2]

1 Frederick Research Center, Nicosia, Cyprus

2 Department of Civil Engineering, Frederick University of Cyprus, Nicosia, Cyprus

*Address all correspondence to: ioangian@outlook.com

References

[1] Fletcher IA, Welch S, Torero J, Carvel R, Usmani A. Behaviour of concrete structures in fires. Journal of Thermal Science. 2007;**11**(2):37-52. DOI: 10.2298/TSCI0702037F

[2] Poljansek M, Nikolova B, Dimova S, Pinto A, Sousa L. Eurocodes – Background and Applications – Structural Fire Design. Worked Examples. Joint Research Centre: Institute for the Protection and Security of the Citizen. Luxembourg: Publications Office of the European Union; 2014. DOI: 10.2788/85432

[3] Katarzyna M, Hager I, Korniejenko K. Material solutions for passive fire protection of buildings and structures and their performances testing. Procedia Engineering. 2016;**151**:284-291. DOI: 10.1016/j.proeng.2016.07.388

[4] Khoury GA. Passive fire protection of concrete structures. Structures and Buildings. 2008;**161**:135-145. DOI: 10.1680/stub.2008.161.3.135

[5] Mizuriaev SA, Zhigulina AY, Mamonov AN, Ganechkina KV. Fire resistant thermal insulation material with regulated moulding viscosity and mixed bonding agent. Procedia Engineering. 2016;**153**:604-608. DOI: 10.1016/j.proeng.2016.08.202

[6] Xu Q, Li GQ, Jiang J, Wan YC. Experimental study of the influence of topcoat on insulation performance of intumescent coatings for steel structures. Fire Safety Journal. 2018;**101**:25-38. DOI: 10.1016/j.firesaf.2018.08.006

[7] Bilotta A, De Silva D, Nigro E. Tests on intumescent paints for fire protection of existing steel structures. Construction and Building Materials. 2016;**121**:410-422. DOI: 10.1016/j.conbuildmat.2016.05.144

[8] Papadopoulos AM. State of the art in thermal insulation materials and aims for future developments. Energy and Buildings. 2004;**37**:77-86. DOI: 10.1016/j.enbuild.2004.05.006

[9] Davidovits J. Geopolymers. Journal of Thermal Analysis. 1991;**37**(8):1633-1656. DOI: 10.1007/BF01912193

[10] Komnitsas K, Zaharaki D. Geopolymerisation: A review and prospects for the minerals industry. Minerals Engineering. 2007;**20**(14):1261-1277. DOI: 10.1016/j.mineng.2007.07.011

[11] Davidovits J. Fire proof geopolymeric cements. In: Proceedings of the 2nd Geopolymere International Conference (GEOPOLYMER '99); 30 June – 2 July 1999; Saint-Quentine. Siant-Quentine: Geopolymer Institute; 2003. pp. 165-169

[12] Zhao R, Sanjayan JG. Geopolymer and Portland cement concretes in simulated fire. Magazine of Concrete Research. 2011;**63**:163-173. DOI: 10.1680/macr.9.00110

[13] Gluth JGJ, Rickard W, Werner S, Pirskawetz S. Acoustic emission and microstructural changes in fly ash geopolymer concretes exposed to simulated fire. Materials and Structures. 2016;**49**(12):5243-5254. DOI: 10.1617/s11527-016-0857-x

[14] Davidovits J. Geopolymer chemistry and properties. In: Proceedings of the 1st Geopolymer International Conference (Geopolymer '88); 1-3 June 1988; Compiegne. Siant-Quentine: Geopolymer Institute; 1997. p. 25-48

[15] Barbosa V, McKenzie K. Thermal behavior of inorganic polymers and composites derives from sodium polysialate. Materials Research Bulletin. 2003;**38**:319-331. DOI: 10.1016/S0025-5408(02)01022-X

[16] Zhang HY, Kodur V, Qi SL, Cao L, Wua B. Development of metakaolin–fly ash based geopolymers for fire resistance applications. Construction and Building Materials. 2014;**55**:38-45. DOI: 10.1016/j.conbuildmat.2014.01.040

[17] Bakharev T. Thermal behaviour of geopolymers prepared using class F fly ash and elevated temperature curing. Cement and Concrete Research. 2006;**36**(6):1134-1147. DOI: 10.1016/j.cemconres.2006.03.022

[18] Guerrieri M, Sanjayan JG. Behaviour of combined fly ash/slag based geopolymers when exposed to high temperatures. Fire and Materials. 2010;**34**:163-175. DOI: 10.1002/fam.1014

[19] Rovnanik P, Bayer P, Rovnanikova P. Characterization of alkali activated slag paste after exposure to high temperatures. Construction and Building Materials. 2013;**47**:1479-1487. DOI: 10.1016/j.conbuildmat.2013.06.070

[20] Cheng TW, Chiu JP. Fire-resistant Geopolymer produced by granulated blast furnace slag. Minerals Engineering. 2003;**16**:205-210. DOI: 10.1016/S0892-6875(03)00008-6

[21] Sakkas KM, Nomikos P, Sofianos A, Panias D. Inorganic polymeric materials for passive fire protection of underground constructions. Journal of Fire Materials. 2013;**37**:140-150. DOI: 10.1002/fam.2119

[22] Rickard W, Vickers L, Van Riessen A. Performance of fibre reinforced, low density matakaolin geopolymers under simulated fire conditions. Applied Clay Science. 2013;**73**:71-77. DOI: 10.1016/j.clay.2012.10.006

[23] Giannopoulou I, Robert PM, Sakkas K, Petrou M, Nicolaides D. High temperature performance of geopolymers based on construction and demolition waste. Journal of Building Engineering. 2023;**72**:106575. DOI: 10.1016/j.jobe.2023.106575

[24] European Commission. Development and Implementation of Initiatives Fostering Investment and Innovation in Construction and Demolition Waste Recycling Infrastructure. Luxembourg: Publications Office of the European Union; 2018. DOI: 10.2873/11837

[25] Menegaki M, Damigos D. A review on current situation and challenges of construction and demolition waste management. Current Opinion in Green and Sustainable Chemistry. 2019;**13**:8-15. DOI: 10.1016/j.cogsc.2018.02.010

[26] Soultana A, Valouma A, Bartzas G, Komnitsas K. Properties of inorganic polymers produced from brick waste and metallurgical slag. Minerals. 2019;**9**:551. DOI: 10.3390/min9090551

[27] Kioupis D, Skaropoulou A, Tsivilis S, Kakali G. Valorization of brick and glass CDWs for the development of geopolymers containing more than 80% of wastes. Minerals. 2020;**10**(8):672. DOI: 10.3390/min10080672

[28] Allahverdi A, Kani EN. 18 - use of construction and demolition waste (CDW) for alkali-activated or geopolymer cements. In: Pacheco-Torgal F, Tam VWY, Labrincha JA, Ding Y, De Brito J, editors. Woodhead Publishing Series in Civil and Structural Engineering, Handbook of Recycled Concrete and Demolition

Waste. Cambridge, UK: Woodhead Publishing Limited; 2013. pp. 439-475. DOI: 10.1533/9780857096906.3.439

[29] Alhawat M, Ashour A, Yildirim G, Aldemir A, Sahmaran M. Properties of geopolymers sourced from construction and demolition waste: A review. Journal of Building Engineering. 2022;**50**:104104. DOI: 10.1016/j.jobe.2022.104104

[30] Luhar S, Cheng TW, Nicolaides D, Luhar I, Panias D, Sakkas K. Valorisation of glass waste for development of Geopolymer composites – Mechanical properties and rheological characteristics: A review. Construction and Building Materials. 2019;**220**:547-564. DOI: 10.1016/j.conbuildmat.2019.06.041

[31] ISO. 834: Fire Resistance Tests – Elements of Buildings Construction. Geneva: International Organization for Standardization; 1999. p. 1999

[32] EFNARC. European Federation of National Associations Representing Producers and Applicators of Specialist Building Products for Concrete: Specification and Guidelines for Testing of Passive Fire Protection for Concrete Tunnels Lining; 2009

[33] Sakkas K, Panias D, Nomikos P, Sofianos A. Comparison of fire resistant Geopolymers for passive fire protection of concrete tunnel linings. Open Access Library Journal. 2017;**4**:e3327. DOI: 10.4236/oalib.1103327

[34] He R, Dai N, Wang Z. Thermal and mechanical properties of Geopolymers exposed to high temperature: A literature review. Advances in Civil Engineering. 2020;Article ID 7532703. DOI: 10.1155/2020/7532703

[35] Bajare D, Vitola L, Dembovska L, Bumanis G. Waste steam porous alkali activated materials for high temperature application. Frontiers of Materials Science. 2019;**6**(92):1-13. DOI: 10.3389/fmats.2019.00092

[36] Lahoti K, Tan KH, Yang EH. A critical review of geopolymer properties for structural fire resistance applications. Construction and Building Materials. 2019;**221**:514-526. DOI: 10.1016/j.conbuildmat.2019.06.076

[37] Gultrone G, Sidraba I, Sebastian E. Mineralogical and physical characterization of the bricks used in the construction of the "Triangul bastion", Riga (Latvia). Applied Clay Science. 2025;**28**:297-308. DOI: 10.1016/j.clay.2004.02.005

Chapter 5

Numerical Simulation of Thermal Shock for Glass-Glass BIPVs in Fire: Gaps and Challenges

Chiara Bedon, Lorenzo Veronese, Riccardo Del Bello, Nicola Cella and Yu Wang

Abstract

Glass-glass photovoltaic (PV) technologies for building integrated (BIPV) applications are increasingly used in construction, due to several advantages. Besides, many complex aspects characterize the performance of these building components, and robust technical knowledge by many experts (with electrical, mechanical, structural, and architectural backgrounds) is usually required for their analysis. In ordinary BIPV applications, for example, thin glass covers are required to sustain thermal loads that derive from ordinary operational conditions, possible superimposed mechanical loads, even under accidents, such as fire. As such, among others, it is important to realistically predict the occurrence of first cracks in glass due to thermal shock, which is frequently followed by fallout. This chapter analyzes the potential of Finite Element (FE) numerical strategies for a more in-depth assessment of glass-glass BIPV performances in fire, to possibly support their optimal design and functionality assessment. Based on the literature background, a standard furnace test is numerically reproduced for a 1.046 m × 1.719 m glass-glass BIPV sample with 6 mm thick tempered covers and 1 mm thick Ethylene Vinyl Acetate (EVA) encapsulant. The attention is focused on the heat transfer transient analysis for first crack and failure time detection due to thermal shock of the exposed glass cover. The simulations give evidence of thermal shock after about 6′ of fire exposure, which fits well with the available experimental evidences. The critical discussion of comparative results, however, shows a major sensitivity to input material properties and poses the basis for further experimental and numerical investigations.

Keywords: glass, photovoltaics (PV), building integrated photovoltaics (BIPV), ethylene vinyl acetate (EVA), fire, thermal shock, modeling, experiments

1. Introduction

PV solutions for building integrated (BIPV) applications represent widely used composite systems, whose multi-functional performances are particularly challenging to address [1, 2].

Among others, the effect of both thermal and mechanical loads is known to represent a very important and often critical aspect for product design, certification, as well as durability and functionality optimization. In these systems, thin glass layers are used for the front cover (or both front and back covers) to protect the solar cells and provide an appropriate load-bearing capacity to the resisting section (**Figure 1(a)**). The realistic estimation of the propagation of first cracks in glass is hence critical to preserve the integrity of the system, both under ordinary or even extreme design actions. Fire accidents, among others, are even more critical because of superimposed thermal and mechanical effects (**Figure 1(b)**), hence resulting in a challenging detection and analysis of resisting and failure mechanisms.

To verify the failure time of PVs in fire, complex experimental tests are commonly carried out on full-scale samples and prototypes, according to existing standards and regulations [3–6]. Experimental investigations can be also extended to large size BIPV facades, which are even more challenging to assess and mitigate [7].

In this chapter, the attention is given to the use of numerical models and Finite Element (FE) methods for the transient heat transfer, thermal analysis of glass-glass BIPVs in fire, and particularly to their possible use for the detection of first damage and crack initiation in the glass components that are directly exposed to fire, as a consequence of heat propagation (thermal shock). To this aim, a major effort is derived from the support of literature documents that are used to collect experimental evidences for full-scale BIPVs in fire conditions [8], to obtain some input properties for the constituent materials [9, 10], and to give feedback for the calibration of FE boundaries and failure detection strategies for glass elements in fire [11].

The presently reported numerical analysis, carried out in ABAQUS/Standard [12], shows that—even rather challenging—the use of FE simulations can provide thermal scenarios and thermal gradient estimates that could be efficiently combined with conventional simplified methods for glass failure detection due to thermal shock. Besides, several influencing parameters should be taken into account, especially the temperature sensitivity of material properties. Moreover, the study suggests that the availability of an adequate number of experiments, as well as a detailed instrumentation for

(a) (b)

Figure 1.
Examples of glass-glass photovoltaics and fire accident.

BIPV experimental samples under fire, would facilitate a sound numerical validation of FE models, at the local and global level.

The present outcomes, in this sense, represent preliminary promising results of a research investigation that will be necessarily further extended to a multitude of configurations (i.e., materials, layout and size aspects, etc.), as well as to more elaborated thermo-mechanical considerations that could support a robust in-depth analysis of the BIPV load-bearing performance and capacity under extreme conditions.

2. Simplified method for detecting thermal shock

Commonly, a simple and practical approach to detect first glass cracking is based on the limitation of thermal gradients that occur in the hottest and coldest regions of glass [13]. Referring to different glass types and edge production details, the conventional values as in **Table 1** can be taken into account for this purpose [14].

Such an approach, while simplified, can be efficiently combined with refined numerical models that can take into account the layered composition of BIPVs. Here, multiple components and layers with different thermo-physical and mechanical properties (as well as sensitivity to temperature variations) are used to assemble these thin sandwich sections.

In this chapter, attention is focused on the potential of FE numerical analysis (in combination with simplified methods as in **Table 1**) to detect the first crack initiation in glass-glass BIPVs in fire, which corresponds to thermal shock. Any other effect of possible mechanical loads (superimposed to thermal loads) is disregarded.

3. Numerical analysis of BIPVs in fire

3.1 Reference BIPV system

For numerical validation and comparative purposes, the full-scale BIPV specimen experimentally investigated in the work of Huang et al. [8] under fire conditions (and originally identified as "Type I BIPV") was taken into account in this study.

		Edges		
	Glass type	**As-Cut or Arrissed**	**Smooth ground**	**Polished**
Allowable thermal gradient [°C]	AN (< 12 mm)	35	40	45
	HS		100	
	FT		200	
Allowable thermal stress [MPa]	AN (< 12 mm)	20.34	23.24	26.15
	HS		58.10	
	FT		116.20	

Table 1.
Allowable thermal gradients ($\Delta_{allowable}$) for glass, and corresponding allowable thermal stresses ($\sigma_{T,allowable}$).

The reference specimen consisted of a standard BIPV sandwich section inclusive of two tempered glass layers (6 mm in thickness), the interposed Ethylene Vinyl Acetate (EVA) encapsulant (1 mm in thickness), and the embedded monocrystalline Si-based solar cells ($0.156 \times 0.156\ m^2$ their size). Transparent glass was used for both the front and back covers. The BIPV module size spanned over $1.046 \times 1.719\ m^2$, with solar cells arranged in a 6×10 array. A standard, continuous aluminum frame was finally used to fix the edges of the BIPV specimen to the test setup (furnace). Refractory silicone and concrete layers were used to fill possible gaps between the frame and the furnace walls and to avoid the leakage of flames or smoke during the fire experiment.

Overall, the test setup was arranged to impose a time-temperature curve on the exposed glass cover that could comply with ISO and CNS standard temperature curves for fire testing [15, 16]. According to Huang et al. [8], the reference test on "Type I BIPV" sample was carried out with the support of 17 uniformly distributed flame burners in the furnace, and interrupted after ≈13′. For monitoring the temperature evolution inside the furnace during the fire test, a total of 16 thermocouples and 4 feedback temperature sensors were used. On the unexposed surface of "Type I BIPV" specimen, an additional set of 7 K-type thermocouples (TC#n) was taken into account. Most importantly, for the present comparative analyses, TC#1 was placed in the centre of glass cover, while TC#13 was installed at the mid-span section of frame mullion.

Notably, the original experiment by Huang et al. [8] was carried out for a BIPV sample under ordinary electrical operational conditions, thus the observed failure mechanisms included the effect of fire exposure and the combined effect of internal heat generated by solar cells, as well as flashover. The first cracks and fallout of the inner glass layer (i.e., inside the furnace) were experimentally observed in the work of Huang et al. [8] after 6′17″ of fire exposure, with an average hoven temperature in the order of ≈600°C. The final catastrophic and explosive failure of glass components was recorded after 12′16″, with a mean furnace temperature up to ≈800°C.

For the presently reported investigation, the attention was focused on the analysis and estimation of thermal shock (i.e., first crack occurrence) in the front glass layer directly exposed to fire, and thus on the comparative analysis of the experimental feedback with the present numerical estimates for the first 7 minutes of fire exposure only (corresponding to a maximum temperature peak of ≈600°C). Notably, a second BIPV specimen was also experimentally investigated in the work of Huang et al. [8], and detected as "Type II BIPV". In this second case, the first cracks and fallout of the exposed glass layer were observed after 6′45″ of fire loading. This second specimen, however, was disregarded for the present numerical analysis, since characterized by different assembly details, in particular in terms of the presence and arrangement of the junction boxes and the electrical components.

3.2 Modeling strategy

The reference numerical model was assembled in ABAQUS/Standard, and consisted of 3D solid, 8-node elements for the fire analysis (DC3D8 type from ABAQUS library). Each constituent layer of the BIPV specimen was numerically reproduced based on nominal geometrical features, including individual parts for the tempered glass layers, the EVA encapsulant with the embedded solar cells, and the aluminum supporting frame (**Figure 2**). To this aim, a major advantage was taken from past modeling experiences for structural glass components in fire [9–11].

A fine mesh pattern was used for the description of BIPV components, including 4 elements in the thickness of each one of them. The maximum edge size of solid

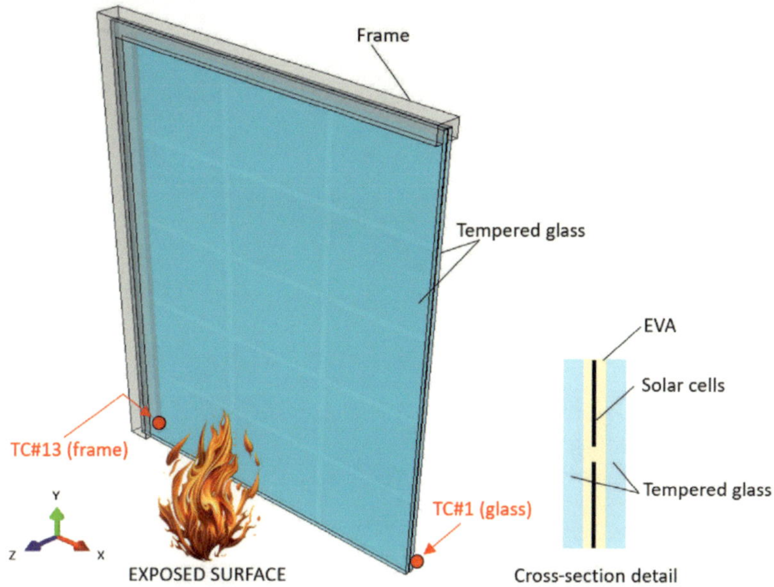

Figure 2.
Axonometry of the reference numerical model (1/4th geometry, ABAQUS).

elements for the BIPV section was fixed at 5 mm. To preserve the computational cost of analyses, one-fourth of the nominal specimen was numerically analysed, with ideal thermal boundary conditions on the symmetry planes and the perimetral surfaces of the frame components (i.e., in contact with the furnace wall and rigidly connected to the laboratory setup). Such a numerical assembly resulted in a total of ≈67,000 elements and ≈94,000 DOFs. **Figure 2** shows a schematic cross-section detail of model components, with evidence of the selected control points for temperature analysis on the unexposed surface of the sample (TC#1 for glass and TC#13 for the frame).

A transient heat transfer analysis was carried out in ABAQUS on the model assembly of **Figure 2**. According to Huang et al. [8], the average ambient temperature of 27°C was taken into account as the initial condition for the numerical analysis. For simulating the fire exposure on the exposed surface of the specimen, the average time-temperature curve derived from the experiment conducted by Huang et al. [8] was taken into account, under uniform exposure assumptions (**Figure 3**). A set of radiation and convection "thermal interactions" was then assigned to the exposed and unexposed surfaces of BIPV, according to Louter et al. [9–11].

Finally, special attention was dedicated to the thermo-physical characterization of materials, whose properties are presented in **Figure 4** and **Table 2**.

In particular, their sensitivity to temperature variations was carefully taken into account, based on a selection of experimental literature data. Glass was in fact numerically described as in the work of Louter et al. [9–11]. Regarding the EVA encapsulant, the input parameters were estimated based on the work of Jia and Zhang [17]. In the description of solar cells, literature input for silicon components was taken into account, such as the work of Desai et al. [18, 19]. For the supporting metal frame, finally, material properties according to Vosteen et al. [20, 21] were taken into account.

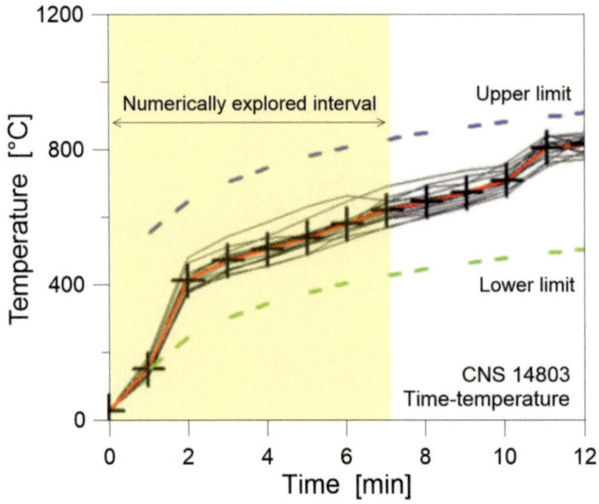

Figure 3.
Average time-temperature curve in use for the numerical analysis of the examined BIPV specimen.

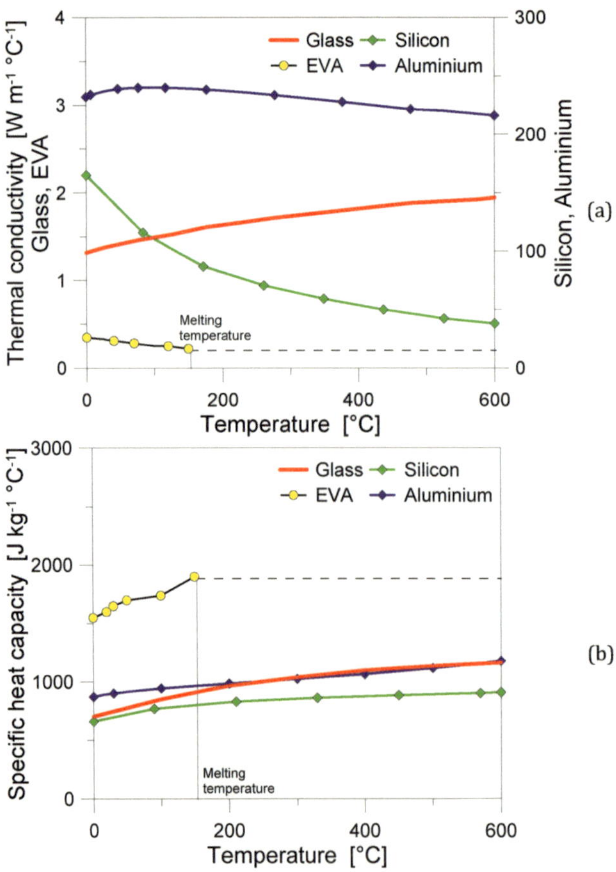

Figure 4.
Input material properties for the heat transfer transient analysis in ABAQUS.

Material	Modulus of elasticity [GPa]	Thermal conductivity [W m^{-1}°C^{-1}]	Specific heat capacity [J kg^{-1}°C^{-1}]	Thermal expansion [1/°C]	Density [kg/m^3]
Glass	70	1.38	715	8.5×10^{-6}	2500
EVA	0.013	0.32	1600	1.6×10^{-4}	960
Silicon	166	140	720	2.62×10^{-6}	2330
Aluminum	70	237	900	2.5×10^{-6}	2700

Table 2.
Reference material properties (room temperature). Thermo-physical values extrapolated from **Figure 4.**

As an important aspect of numerical analysis, the examined temperature range for the purpose of the present study was limited to 0–600°C, and the description of material properties was carried out according to **Figure 4**. Additional attention was numerically given to EVA, however, to possibly assess the numerical effects of its sensitivity to temperature increase, compared to the other constituent materials. The EVA encapsulants for PV applications are in fact characterized by thermo-chemical processes that can affect their performance. Discolouration, acetic acid formation, VA content variation, polymer chain conjugation, corrosion of metallization, and ability to absorb moisture are the main problems of EVA encapsulants [22]. Crystallization and softening processes usually take place for maximum temperatures around 80–100°C [23, 24], which is relatively low compared to fire scenarios or event hotspots for pre-stressed glass-glass BIPV systems [24]. Following MatWeb [25] and due to major uncertainty in such a material characterization, in the absence of more specific data, an average value of ≈150°C was tentatively taken into account in the present numerical study, as the possible melting temperature for the EVA layer, which is considerably lower than the other BIPV materials (≥ 550°C). This last aspect was preliminarily assessed by comparing two different numerical model calibrations. In the reference FE assembly, the thermo-physical features of EVA as in **Figure 4** were in fact kept constant for temperatures higher than 150°C. For comparative purposes, a second simulation was carried out with the same material properties of **Figure 4**, but accounting for possible EVA degradation and instability, that is by disregarding any thermo-physical contribution of EVA layers for temperatures higher than 150°C. Such an approach, while roughly approximate, was herein considered for emphasizing the contribution of EVA encapsulant in the thermal response of the examined BIPV system.

4. Results

4.1 Validation

The typical thermal numerical results were preliminarily post-processed to verify the consistency and accuracy of model features and calibration, towards the experimental data reported in the work of Huang et al. [8]. The major limitation, in such a comparative stage, was represented by limited experimental data/sensors on the BIPV specimen. In this sense, it is important to highlight that the experimental temperature trend in time on the unexposed glass surface was found mostly uniform from the available thermocouples.

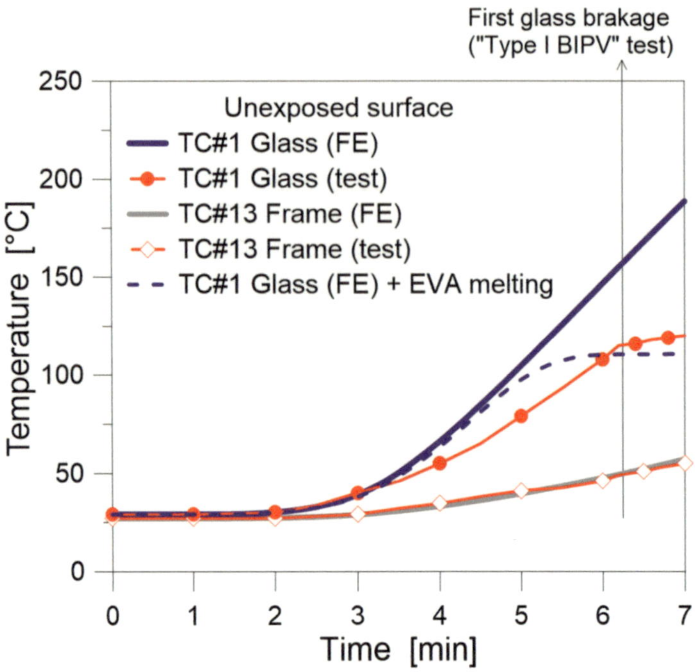

Figure 5.
Temperature evolution on the unexposed surface of the examined BIPV specimen.

Figure 5, in this regard, shows the typical evolution of temperature on the unexposed surface of the BIPV sample ("Type I"), for the two previously selected control points of **Figure 2** (i.e., TC#1 in the centre of the glass back cover and TC#13 at the mid-span section of the aluminum mullion). These two temperature-time experimental curves are compared to the presently obtained numerical predictions, both for the glass cover and the frame. The experimental occurrence of the first glass breakage (on the exposed surface) is also emphasized in **Figure 5**.

Most importantly, the numerical results for glass temperature are presented by accounting for possible EVA degradation with high temperatures, as described in Section 3. Such a modeling approach is extremely simplified but can provide support for the elaboration of future studies.

In terms of temperature trends on the frame (on the unexposed BIPV surface), it can be seen from **Figure 5** that the present FE model fits very well the corresponding measurements in time, from the furnace test. At the time of failure, the recorded temperature on the frame was estimated in the order of ≈50°C, which can be still associated with limited fire consequences and a thermo-physical and mechanical response of the frame itself like at room/ambient temperature (i.e., **Figure 4**). Most importantly, however, major effects derive from the coupled response of the frame with the BIPV section, which typically manifests in major thermal bridge effects for the glass.

For the glass cover on the unexposed side of the BIPV sample, as shown in **Figure 5**, much higher temperature values were in fact predicted, both numerically and experimentally. Moreover, an important effect of the EVA encapsulant can also be perceived from the proposed numerical results, thus suggesting the high complexity and uncertainty of such a thermo-mechanical topic and the need for further extended investigations. From a more detailed analysis of a few comparative data, it can be seen, for example, that the partial overestimation of the experimental temperatures, based on the present modeling strategy, starts from about 3′ of fire exposure. Notably, such an effect could derive from a wide series of aspects, both on the experimental and on the numerical side. The availability of rather scarce experimental results from the furnace setup, however, as well as the absence of specific material properties for "Type I BIPV" components, makes particularly difficult a more in-depth analysis, justification, and optimal calibration of the presently adopted numerical modeling assumptions. Interestingly, **Figure 5** shows that when material degradation is considered for the EVA encapsulant, the measured temperature on the exposed glass surface presents a marked drop at about 5′ of fire exposure, which recalls the experimental trend in time. Anyway, additional studies, inclusive of experimental analyses on BIPV materials and components, are necessarily required.

Finally, it is important to observe that at the experimental time of the first glass breakage for the examined BIPV sample, the average temperature in the furnace was measured at about ≈600°C, where the latter represents an average hoven temperature only. Notably, in alignment with the average CNS time-temperature curve of **Figure 3**, the numerical analysis resulted in a furnace temperature of ≈585°C after 6′17″ of fire exposure, which fits well with the experimental evidences of Huang et al. [8]. At the same instant, the numerical temperature in the BIPV components was found highly sensitive to the presence of the frame and the embedded solar cells, with a maximum peak of about ≈430°C on the exposed surface (**Figure 6**).

4.2 Glass failure detection

According to Section 2 and **Figure 6**, it is clear that the temperature distribution in FE components is of primary importance for failure detection, based on simplified methods. Accordingly, a preliminary attempt was carried out to predict the corresponding thermal gradient and compare it to limit values of **Table 1**. The calculation was numerically carried out by considering—both on the exposed and unexposed

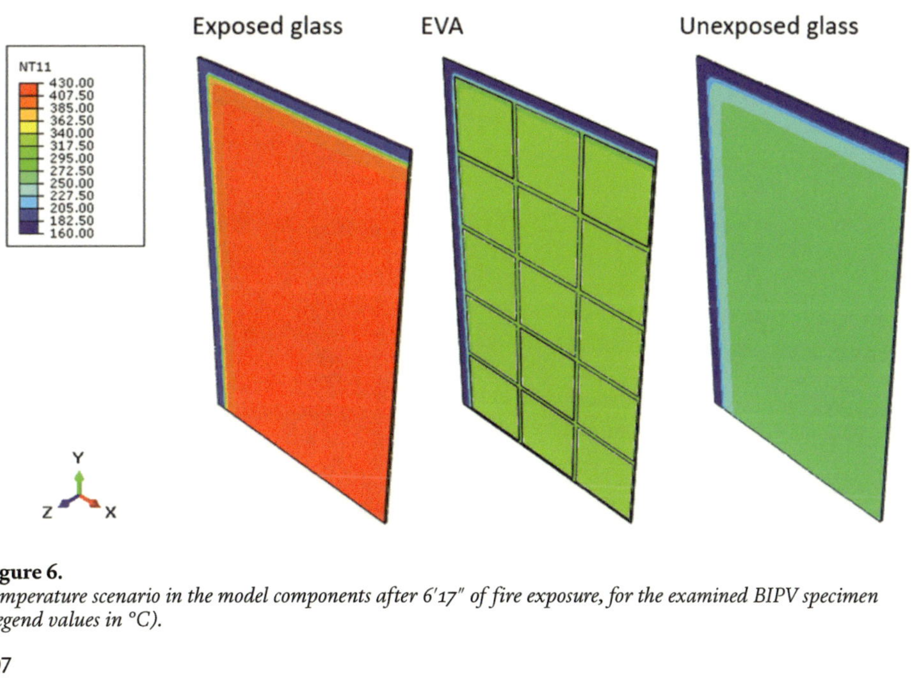

Figure 6.
Temperature scenario in the model components after 6′17″ of fire exposure, for the examined BIPV specimen (legend values in °C).

surfaces of glass—a control point in the centre of glass and a control point at the top glass corner (behind the frame). The so-obtained results are proposed in **Figure 7**.

It is important to note that due to the lack of thermal gradient experimental evidences for the selected control points and time interval, some numerical considerations can be only carried out on the basis of the earlier validation of **Figure 5**. As shown, the present modeling assumptions resulted in thermal gradients that rapidly increase with time. For the unexposed glass layer, this gradient was quantified at about ≈100° after 7′ of exposure, which is still associated with safe/non-critical temperature ranges for tempered glass (**Table 1**). In the case of the exposed glass panel, however, the conventional limit value of 200°C for tempered glass was numerically exceeded after 5′47″ of fire exposure, which represents a less than ≈30″ underestimation, compared to the reference test for "Type I BIPV" sample.

Such a quantified scatter, which is on the conservative side, could derive from possible uncertainties on the calibration of material properties (i.e., that are currently based on input features from literature) that do not reflect the thermo-physical and mechanical properties of the tested BIPV sample [26]. This is especially the case of EVA encapsulant and its possible thermal instability with temperatures higher than 150°C (**Figure 4**). Another justification for such a scatter for the expected failure time due to thermal shock could derive from the real fire exposure of the BIPV specimen during the furnace test, which was numerically assumed as perfectly uniform, over the whole exposed surface of the FE model, but in real systems can be affected by some variations (i.e., **Figure 3**), both in the elevation and in the width specimens [27].

4.3 Temperature effects on material properties

The effect of material sensitivity to the progressive temperature modification was finally pointed out. The comparative results showed that disregarding the

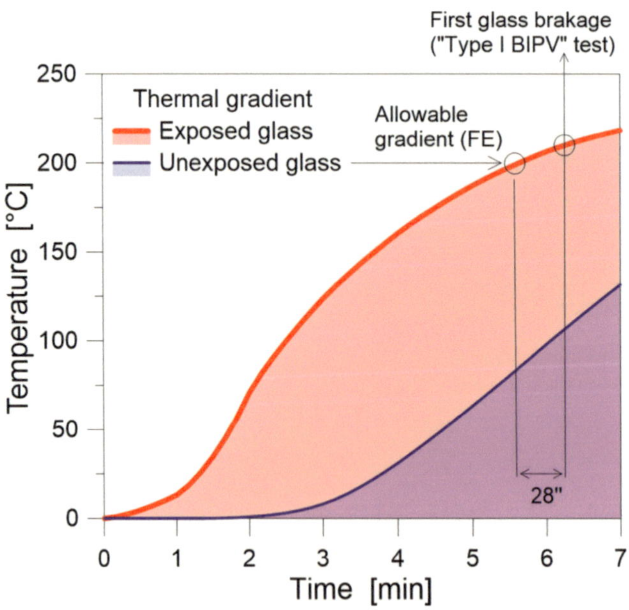

Figure 7.
Numerically predicted thermal gradient, and simplified detection of glass failure on the exposed surface of the examined BIPV specimen.

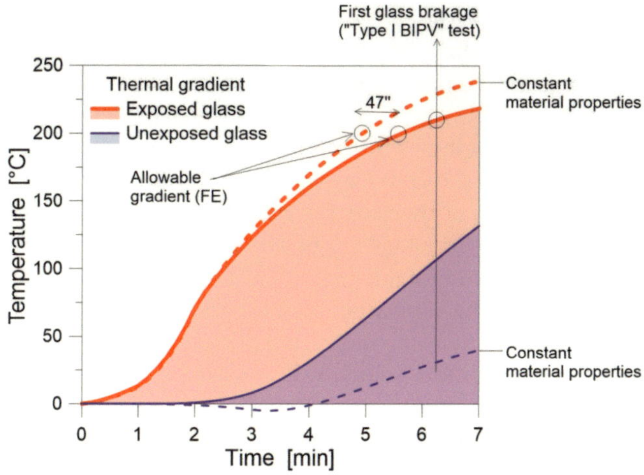

Figure 8.
Effect of input material properties on the numerically predicted thermal gradient and simplified detection of glass failure on the exposed surface of the examined BIPV specimen.

modification of material parameters with the increasing temperature can have major effects for the transient analysis of similar systems, especially for glass failure detection. Typical results can be seen in **Figure 8**.

In the case of exposed glass panel, the numerically predicted thermal gradient was found to increase with constant material properties, compared to the reference model. Such an effect—while on the safe side—underestimates the load-bearing capacity and can have major consequences, especially in terms of mechanical performance assessment in fire. For the unexposed glass layer, even more pronounced effects were observed in terms of corresponding thermal gradients. Also in this case, the reason derives from the progressive heat propagation in the BIPV thickness, where multiple components (glass, EVA, cells) contribute to the thermal scenario on the unexposed surface.

5. Conclusions

The thermo-mechanical performance of photovoltaic (PV) and building integrated photovoltaics (BIPV) is a well-known challenging task, given that multiple aspects and technical knowledges should be properly taken into account, on the electrical, engineering, and even architectural side.

Fire safety issues are particularly challenging and crucial for BIPVs, where thin composite sections are assembled with glass covers, EVA encapsulants, and embedded solar cells. The integrity and functionality of BIPVs are strictly related to the load-bearing capacity of glass covers, which are commonly subjected to a combination of thermal and mechanical loads.

In this chapter, the use of Finite Element (FE) numerical models, in combination with simplified strategies of literature, was taken into account to preliminarily assess their potential in predicting the initiation of first cracks in the exposed glass layer, due to thermal shock.

The comparative analysis for a full-scale BIPV specimen in a standard furnace test setup of literature was based on original numerical simulations carried out in ABAQUS/Standard. A full 3D solid model was described, in which all the BIPV constituent components were properly reproduced based on nominal geometrical features and literature material properties.

The analysis of numerical results towards literature experimental data showed that—even extremely challenging—the estimation of failure time and first crack initiation in the exposed glass cover can be somehow captured by numerical simulations. Besides, the description of thermal and mechanical boundary conditions for the reference experimental setup may have major effects on numerical results and should be possibly supported by much more details as possible. The number and position of sensors and instruments for the experimental analysis, moreover, should be in general maximized to facilitate both the local and global comparative analysis of numerical estimates, in order to obtain a more robust validation of simulation outcomes. Among the major simplifications of the present numerical investigations, a uniformly distributed, average time-temperature curve for simulating the fire exposure of the BIPV specimen was taken into account, and this assumption can be responsible for a not fully realistic description of the thermal scenario inside the test furnace. Major gaps in the prediction of thermal shock and failure time for the exposed glass panel may also derive from the detailed calibration of the constituent materials, particularly their sensitivity and possible degradation with increasing temperatures.

In any case, the present preliminary study showed promising results and a good correlation (on the safe side) of the expected failure time, for the examined BIPV experiment. Further studies will be carried out by taking into account multiple geometrical and mechanical configurations of practical interest, as well as fire scenarios, that will be investigated both experimentally and numerically, to derive possible guidelines and practical rules for the calibration of efficient FE modeling strategies.

Acknowledgements

This research study was carried out in the framework of the ongoing Particular Relevance Italy-China bilateral project "3FiRES - Research on BIPV Photovoltaic Facades for Fire Spread Mechanisms, Structural Failures and Resilience Improvement Methodologies" (2024–2025). The work was supported by the National Key R&D Program of China (grant number 2023YFE0116700) and the Italian Ministry of Foreign Affairs and International Cooperation (grant number CN24GR03).

Conflict of interest

The authors declare no conflict of interest.

Author details

Chiara Bedon[1*], Lorenzo Veronese[1], Riccardo Del Bello[1], Nicola Cella[1] and Yu Wang[2]

1 Department of Engineering and Architecture, University of Trieste, Italy

2 State Key Laboratory of Fire Science, University of Science and Technology of China, Hefei, P. R. China

*Address all correspondence to: chiara.bedon@dia.units.it

References

[1] Maghrabie HM, Abdelkareem MA, Al-Alami AH, Ramadan M, Mushtaha E, Wilberforce T, et al. State-of-the art technologies for building-integrated photovoltaic systems. Buildings. 2021;**11**:383

[2] Khalifa Abojela ZR, Desa MKM, Sabry AH. Current prospects of building-integrated solar PV systems and the application of bifacial PVs. Frontiers in Energy Research. 2023;**11**:1164494

[3] Yang R, Zang Y, Yang J, Wakefield R, Nguyen K, Shi L, et al. Fire safety requirements for building integrated photovoltaics (BIPV): A cross-country comparison. Renewable and Sustainable Energy Reviews. 2023;**173**:113112

[4] Despinasse M-C, Krueger S. First developments of a new test to evaluate the fire behavior of photovoltaic modules on roofs. Fire Safety Journal. 2015;**71**:49-57

[5] Stølen R, Li T, Wingdahl T, Steen-Hansen A. Large- and small-scale fire test of a building integrated photovoltaic (BIPV) façade system. Fire Safety Journal. 2024;**144**:104083

[6] Wang Y, Xiao C, Bedon C. Performance of photovoltaic panels with different inclinations under uniform thermal loading. International Journal of Thermal Sciences. 2025;**208**:109489

[7] Bedon C, Zhang X, Santos F, Honfi D, Kozlowski M, Arrigoni M, et al. Performance of structural glass facades under extreme loads - Design methods, existing research, current issues and trends. Construction and Building Materials. 2018;**163**:921-937

[8] Huang Y-C, Lee S-K, Chan C-C, Wang S-J. Full-scale evaluation of fire-resistant building integrated photovoltaic systems with different installation positions of junction boxes. Indoor and Built Environment. 2017;7(9):1259-1271. DOI: 10.1177/1420326X17713256

[9] Louter C, Bedon C, Kozlowski M, Nussbaumer A. Structural response of fire-exposed laminated glass beams under sustained loads; exploratory experiments and FE-simulations. Fire Safety Journal. 2021;**123**:103353

[10] Bedon C, Louter C. Thermo-mechanical numerical analyses in support of fire endurance assessment of ordinary soda-lime structural glass elements. Journal of Structural Fire Engineering. 2023;**14**(4):522-546

[11] Kozlowski M, Bedon C. Sensitivity to input parameters of failure detection methods for out-of-plane loaded glass panels in fire. Fire. 2021;**4**(1):5. DOI: 10.3390/fire4010005

[12] Simulia (2024). Abaqus Computer Software. Providence, RI, USA;

[13] Galuppi L, Franco A, Bedon C. Architectural glass under climatic actions and fire: Review of state of the art, open problems and future perspectives. Buildings. 2023;**13**:939

[14] prEN Thstr: 2004. Glass in Building: Thermal Stress Calculation Method. European Committee for Standardization (CEN): Brussels, Belgium; 2004

[15] ISO 834-1: 1999. Fire-Resistance Tests – Elements of Building Construction – Part 1: General Requirements. Geneva: International Organization for Standardization; 1999

[16] CNS 14803-2010. Method of Fire Resistance Tests for Rolling Shutter of

Buildings. Taipei: Bureau of Standards, Metrology & Inspection, M.C.E.A., ROC; 2010

[17] Jia Y, Zhang J. Thermal conductivity of ethylene-vinyl acetate copolymers with different vinyl acetate contents dependent on temperature and crystallinity. Thermochimica Acta. 2022;**708**:179141

[18] Desai PD. Thermodynamic properties of iron and silicon. Journal of Physical and Chemical Reference Data. 1985;**15**(3):967-983

[19] Morris RG, Hust JG. Thermal conductivity measurements of silicon from 30° to 425°C. Physical Review. 1961;**14**(5):1426-1430

[20] Vosteen LF. Effect of Temperature on Dynamic Modulus of Elasticity of some Structural Alloys. Washington: NACA TN 4348, National Advisory Committee of Aeronautics, Technical Note 4348; 1958

[21] Doležel I, Karban P, Kropík P, Pánek D. Accurate control of position by induction-heating produced thermoelasticity. IEEE Transactions on Magnetics. 2010;**46**:2888-2891

[22] Adothu B, Bhatt P, Kartikay P, Zele S, Costa FR, Oderkerk J, et al. Determination of crystallinity and thermal stability of newly developed thermoplastic polyolefin encapsulant for the c-Si PV module application. In: 2019 IEEE 46th Photovoltaic Specialists Conference (PVSC), Chicago, IL, USA, 2019. 2019. pp. 487-490. DOI: 10.1109/PVSC40753.2019.8980526

[23] Stark W, Jaunich M, Bohmeyer W, Lange K. Investigation of the crosslinking behaviour of ethylene vinyl acetate (EVA) for solar cell encapsulation by rheology and ultrasound. Polymer Testing. 2012;**31**:904-908

[24] Afridi M, Kumar A, Mahmood F, Tamizhmani G. Hotspot testing of glass/backsheet and glass/glass PV modules pre-stressed in extended thermal cycling. Solar Energy. 2023;**249**:467-475

[25] MatWeb – Material Property Data. Overview of Materials for Ethylene Vinyl Acetate Copolymer (EVA), Film Grade. Blacksburg, USA: MatWeb, LLC; 2025. Available from: https://www.matweb.com/search/datasheet.aspx?matguid=f34f846b98f1466780dfb3c0e5c0ca82&ckck=1

[26] Bedon C. Structural glass systems under fire: Overview of design issues - Experimental research and developments. Advances in Civil Engineering. 2017;**2017**:1-18

[27] Vedrtnam A, Bedon C, Youssef MA, Chaturvedi S. Effect of non-uniform temperature exposure on the out-of-plane bending performance of ordinary laminated glass panels. Composite Structures. 2021;**275**:114517